Hierarchy, Information and Power

This book is a collection of selected papers presented at the 2012 annual meeting of the Association of American Geographers in New York honoring James O. Wheeler (1938–2010). The eight papers are informed and inspired by James O. Wheeler's many contributions to urban geography, particularly in the areas of urban hierarchy, information flows, cities in the telecommunications age, and cities as corporate command and control centers. They adopt and extend Jim Wheeler's corporate and/or hierarchical approaches to discuss institutional investment in the U.S., corporate interlocking directorates and fast-growing firms in Canada, corporate intangible assets in South Korea, urban development in Beijing and Macau, and social and cultural diversity of global cities such as New York. Although these two approaches are not the fanciest ones in today's urban geography, they are essential to the understanding of how urban areas are connected and what drives this interconnectedness in this age of globalization.

This book was previously published as a special issue of *Urban Geography*.

Hongmian Gong is Professor of Geography at Hunter College and Professor of Earth and Environmental Sciences at the Graduate Center of the City University of New York, USA. Her research interests are urban service economies in the U.S. and China, and GIS/GPS/mobile phone applications in urban transportation.

Hierarchy, Information and Power

Cities as corporate command and control centers

Edited by

Hongmian Gong

Routledge
Taylor & Francis Group
LONDON AND NEW YORK

First published 2017
by Routledge
2 Park Square, Milton Park, Abingdon, Oxfordshire OX14 4RN
711 Third Avenue, New York, NY 10017

Routledge is an imprint of the Taylor & Francis Group, an informa business

First issued in paperback 2018

British Library Cataloguing in Publication Data
A catalogue record for this book is available from the British Library

ISBN 13: 978-1-138-68996-1 (hbk)
ISBN 13: 978-0-367-02631-8 (pbk)

Typeset in Times New Roman
by RefineCatch Limited, Bungay, Suffolk

Publisher's Note
The publisher accepts responsibility for any inconsistencies that may have arisen during the conversion of this book from journal articles to book chapters, namely the possible inclusion of journal terminology.

Contents

Contents

Citation Information

The chapters in this book were originally published in *Urban Geography*, volume 36, issue 6 (September 2015). When citing this material, please use the original page numbering for each article, as follows:

Chapter 1
Introduction: Hierarchy, information, and power: cities as corporate command and control centers. Special issue in honor of James O. Wheeler
Hongmian Gong
Urban Geography, volume 36, issue 6 (September 2015) pp. 795–797

Chapter 2
The economic geography of institutional investment in the United States, 2010
Milford Green, Sean O'Hagan and Martin Lefebvre
Urban Geography, volume 36, issue 6 (September 2015) pp. 798–822

Chapter 3
The geography of Canadian interlocking directorates: how do they relate to brain circulation?
Sean B. O'Hagan and Murray D. Rice
Urban Geography, volume 36, issue 6 (September 2015) pp. 823–843

Chapter 4
Fast-growing firms as elements of change in Canada's headquarters city system
Murray D. Rice, Donald I. Lyons and Sean B. O'Hagan
Urban Geography, volume 36, issue 6 (September 2015) pp. 844–863

Chapter 5
Interaction of corporate and urban systems: accumulation of intangible assets
Sam Ock Park
Urban Geography, volume 36, issue 6 (September 2015) pp. 864–882

Chapter 6
Macau's role as a recreation/tourist center in the Pearl River Delta city-region
Clifton W. Pannell and Philip H. Loughlin
Urban Geography, volume 36, issue 6 (September 2015) pp. 883–904

Chapter 7
Planning Beijing: socialist city, transitional city, and global city
Chaolin Gu, Yehua Dennis Wei and Ian G. Cook
Urban Geography, volume 36, issue 6 (September 2015) pp. 905–926

Chapter 8
Global cities, cosmopolitanism, and geographies of tolerance
Barney Warf
Urban Geography, volume 36, issue 6 (September 2015) pp. 927–946

Chapter 9
A method for delineating a hierarchically networked structure of urban landscape
Yichun Xie and Ting Ma
Urban Geography, volume 36, issue 6 (September 2015) pp. 947–963

Notes on Contributors

Ian G. Cook is Emeritus Professor in the School of Humanities and Social Science, Liverpool John Moores University, Liverpool, UK.

Hongmian Gong is Professor of Geography at Hunter College and Professor of Earth and Environmental Sciences at the Graduate Center of the City University of New York, New York, USA.

Milford Green is Professor in the Department of Geography, Western University, London, Canada.

Chaolin Gu is Professor in the Department of Urban Planning, Tsinghua University, Beijing, China.

Martin Lefebvre is a PhD candidate in the Department of Geography, Western University, London, Canada.

Philip H. Loughlin is a Researcher in the Department of Economics, Maryville University, Cambridge, USA.

Donald I. Lyons is Professor and Head of the Department of Geography, University College Cork, Cork, Ireland.

Ting Ma is a Researcher in the Institute of Geographic Sciences and Natural Resources Research, Chinese Academy of Sciences, Beijing, China.

Sean B. O'Hagan is Associate Professor in the Department of Geography, Nipissing University, North Bay, Canada.

Clifton W. Pannell is Emeritus Professor in the Department of Geography, University of Georgia, Athens, USA.

Sam Ock Park is Professor Emeritus in the Department of Geography, Seoul National University, Seoul, Korea and Chair Professor of the Department of Public Administration, Gachon University, Seongnam City, Korea.

Murray D. Rice is Associate Professor in the Department of Geography, University of North Texas, Denton, USA.

Barney Warf is Professor in the Department of Geography, University of Kansas, Lawrence, USA.

Yehua Dennis Wei is Professor in the Department of Geography, University of Utah, Salt Lake City, USA.

Yichun Xie is Professor in the Department of Geography and Geology, Eastern Michigan University, Ypsilanti, USA.

INTRODUCTION

Hierarchy, information, and power: cities as corporate command and control centers

Hongmian Gong

Department of Geography, Hunter College of the City University of New York, New York, NY, USA

This collection of papers was originally among several presented in the 2012 Annual Meeting of the Association of American Geographers in New York honoring James O. Wheeler (1938–2010). Jim Wheeler leaves an extensive legacy to geography that can be seen in his many roles within the discipline, including service as teacher, administrator, editor, board member, researcher, and advisor to a host of students and colleagues. Particularly, he was a founding editor of *Urban Geography*, for which he co-edited for more than two decades and contributed 16 editorials.

This special issue focuses on research that has been informed and inspired by Jim Wheeler's many contributions to urban geography, particularly in the areas of urban hierarchy, information flows, cities in the telecommunications age, and cities as corporate command and control centers. The eight papers in the special issue adopt and extend Jim's corporate and/or hierarchical approach to understand the contemporary urban changes in the globalization age.

1. Corporate approach

One of Jim's most prominent areas of contribution came through his many published works related to the field of corporate geography, which accounted for over one third of his lifetime journal output. Jim's broad range of interests related to corporate geography included the location of corporate headquarters (Wheeler, 1987, 1988), the geography of finance and investment (Wheeler, 1986), the distribution of fast-growing private companies (Wheeler, 1990a), and location decisions in manufacturing (Wheeler, 1981).

The first four papers in this special issue discuss institutional investment (by Green et al.) in the United States, corporate interlocking directorates (by O'Hagan and Rice) and fast-growing firms (by Rice et al.) in Canada, and corporate intangible assets in South Korea (by Park). Similar to many of Jim's studies, these four papers use data at the corporate level to study the location choices of these firms in cities. However, in addition to strengthening Jim's emphases on corporate headquarters and corporate services, these papers extend Jim's corporate approach to include aspects such as interlocking directorates (by O'Hagan and Rice) and intangible assets of brand values and cultural bundles (by Park).

2. Hierarchical approach

Another one of Jim's most prominent areas of contribution relates to the United States urban/metropolitan hierarchy and its role in information flows, in corporate headquarters and financial services, and in the service economy. Examples of his publications in this area include his seminal papers of "Corporate Spatial Links with Financial Institutions: The Role of the Metropolitan Hierarchy" (Wheeler, 1986), "Information Flows among Major Metropolitan Areas in the United States" (Wheeler & Mitchelson, 1989a), and "The Flow of Information in a Global Economy: The Role of the American Urban System in the 1990" (Mitchelson & Wheeler, 1994). Jim also used the hierarchical approach to study individual cities such as New York and Atlanta, as in "The Corporate Role of New York in the Metropolitan Hierarchy" (Wheeler, 1990b) and "Atlanta's Role as an Information Center: Intermetropolitan Spatial Links" (Wheeler & Mitchelson, 1989b).

This hierarchical approach is used in most papers in this special issue and is updated in the last four papers with international and global perspectives. In addition to the paper by Park applying the corporate and hierarchical approaches to South Korea, the papers by Pannell and Loughlin, Xie and Ma, and Gu et al. use the hierarchical approach for studies outside of North America, in industrializing regions such as China. These latter three papers research the hierarchical interaction and specialization not in a national urban system, but among urban and suburban centers within the Beijing metropolitan area (by Xie and Ma and by Gu et al.) or among cities in the Pearl River Delta city-region (by Pannell and Loughlin). Xie and Ma also noted that Jim was an early adopter of quantitative analysis and GIS. They use GIS and remote sensing techniques to establish a hierarchical urban structure delineation method. Adding to Jim's work on economic dominance of global cities such as New York, Warf discusses in his paper the social and cultural diversity of global cities. These papers complement and broaden Jim's research to reflect the pluralism characteristics (The Editors, 2015) of urban geography.

This special issue honors James O. Wheeler, who is remembered as a true master in using either corporate or hierarchical approach alone, and particularly in using them in seamless combination in urban geographical research. Although these two approaches are not dominant in today's urban geography, they are essential to the understanding of how urban areas are connected and what drives this interconnectedness in the globalization age.

Acknowledgements

I would like to thank the authors for their contribution to this special issue and the anonymous reviewers who graciously provided their service in the review process. Thanks also go to those who presented and/or participated in the three sessions honoring James O. Wheeler at the 2012 Annual Meeting of the Association of American Geographers.

References

The Editors. (2015). Editorial: New urban geographies. *Urban Geography*, *36*, 327–339.

Mitchelson, Ronald A., & Wheeler, James O. (1994). The flow of information in a global economy: The role of the American urban system in 1990. *Annals of the Association of American Geographers, 84*, 87–107.

Wheeler, James O. (1981). Effects of geographical scale on location decisions in manufacturing: The Atlanta example. *Economic Geography*, *57*, 134–145.

Wheeler, James O. (1987). *Fortune* firms and the fortunes of their headquarters metropolises. *Geografiska Annaler, Series B*, *68*, 65–71.

Wheeler, James O. (1988). Spatial ownership links of major corporations: The Dallas and Pittsburgh examples. *Economic Geography, 64*, 1–16.
Wheeler, James O. (1990b). The corporate role of New York in the metropolitan hierarchy. *Geographical Review, 80*, 370–381.
Wheeler, James O., & Mitchelson, Ronald L. (1989a). Information flows among major metropolitan areas in the United States. *Annals of the Association of American Geographers, 79*, 523–543.
Wheeler, James O. (1986). Corporate spatial links with financial institutions: The role of the metropolitan hierarchy. *Annals of the Association of American Geographers, 76*, 262–274.
Wheeler, James O. (1990a). The new corporate landscape: America's fastest growing private companies. *The Professional Geographer, 42*, 433–444.
Wheeler, James O., & Mitchelson, Ronald L. (1989b). Atlanta's role as an information center: Intermetropolitan spatial links. *The Professional Geographer, 41*, 162–172.

The economic geography of institutional investment in the United States, 2010

Milford Green[a], Sean O'Hagan[b] and Martin Lefebvre[a]

[a]Department of Geography, Western University, London, Canada; [b]Department of Geography, Nipissing University, North Bay, Canada

The economic geography of United States' institutional investments as of December 2010 is presented. With a self-declared value of almost $13 trillion, it is the largest class of equity investors. Included are all investors who file Securities Exchange Commission (SEC) form 13f, meaning they manage at least $100 million. Analysis is undertaken at both the firm and metropolitan levels. The largest firms and targets of investment are determined. Geographic concentrations of both investors and their major stock holdings occur in a small subset (the top 10 largest metropolitan areas). The rates of returns on investments are shown to differ by metropolitan area of investment and by urban rank in the urban hierarchy. Intra-metropolitan investment provides worse rates of returns, countering suggestions of informational advantages for local investment. The network of interurban investments shows a clear hierarchical structure, with New York City being the dominant investment centre.

Introduction

Experiences from the latest financial crisis have made the public more aware than ever of the role of large investors in the North American financial system. During the panic phase of the crisis, the focus of attention was on the bankers' culpability and on the incompetence of regulators. Furthermore, words such as institutional investors and hedge funds were incessantly used by media reporting on the situation, with little content attached to them (Smith, 2010). Moreover, the discussion was strictly linked to the financial aspects of investing—gains and loss on stocks and derivatives, the use of complex financial instruments such as collateralized debt obligations (CDOs) and swaps (two parties lend to each other on different terms). Little to no discussion has taken place about the role that geography plays in analysing these investments.

Pursuant to this limitation, the purpose of this article is to investigate whether space matters for institutional investment at a time when competitive forces have significantly internationalized operations in many industries. Clark (2005) argues that social scientists, and economic geographers particularly, continue to conduct research based on the twentieth-century industrial capitalism rather than on the twenty-first-century financial capitalism. Using traditional geographic tools, such as large-scale empiricism, and modern geographic testing, such as log-linear analysis, this research revisits the neglected field to examine two main hypotheses. Does this industry reveal a decentralization of operations or is it highly concentrated with a fair degree of spatial inertia? Furthermore, how are the components of the institutional investment industry (the institutional investors, the investees and their host cities) connected?

Institutional investors are parties with large amounts of capital derived from other peoples' savings to invest. Their role in the economy is to act as highly specialized investors for others. They offer some risk pooling, as individual investment is collectively invested with that of others under the direction of specialist managers. Because institutional investors generally have the freedom to buy and sell large blocks of shares, this type of investor has a great deal of influence in the management of corporations. The ability to influence the selection of directors is not uncommon. Sometimes they can influence which companies remain solvent and which go bankrupt.

Short history of some industrial components

The origin and history of the institutional investment industry of the United States are the stories of several different types of financial institutions. To appreciate the major components of institutional investors better, a short history of each is provided. Definitions of the components of the industry normally include public insurance companies, close-ended mutual funds, pension plans, hedge funds, sovereign wealth funds, endowment funds and unit investment trusts (UITs). Unfortunately, the 13f filings do not show the type of investor firm being recorded.

Investment banks

Investment banks have been involved in institutional investment from the beginning. Sobel (1968, p. 59) provides a simple definition of investment bankers as "bankers who serve companies and institutions rather than the general public". With the institution of the Glass–Steagall Act (active from 1933 to 1999), banks were categorized into one of three classes: investment, commercial and retailing. Prior to the act, the division between different sorts of banking was murky at best. In terms of marketable securities, such as stocks and bonds, investment bankers were deal makers rather than deal takers. They were instrumental in the creation of investment trusts, the forerunners of modern institutional investors, although such trusts were far more speculative in nature. Before the crash of 1929, they were known to be involved in the manipulation and "rigging" of the stock market (Sobel, 1968).

The 1920s was the birth date of modern institutional investment. The banking sector started to woo wealthy individuals with the intent of attracting deposits and provide investment services. Private banks in particular were interested in increasing their capital base to further their market speculations. This was aided by the regulatory regime of the time. As Geisst (1997, p. 120) states:

> The only real constraint was geography. Banks were not able to cross state lines and, in many states such as New York, were not able to cross city and county lines. As a result, the most successful of them were concentrated in New York City and, to lesser extent in Chicago. Their power derived from the connections they had forged over the years with businesses and corporations.

This geographical constraint led to the concentration of the institutional investment firms within a select group of cities headed by New York City.

Investment and private banks currently provide advice, custodial services for securities and facilitate mergers and acquisitions. Since the repeal of the Glass–Steagall Act in 1999, investment banks have again blurred the lines between bank types, often participating in all three types of banking activity. Although investment banks no longer dominate as they once did, they are still among the most important institutional investment firms. The condition of the current banking sector is reminiscent of the banking structure of the 1920s.

Mutual funds

Mutual funds are also a very important form of institutional investment with 13,045 funds with $13.0 trillion in net assets in 2012 (Investment Company Institute [ICI], 2013, p. 9). Of the three main types, open-ended, exchange-traded funds (ETFs) and UITs, only ETFs and UITs are reported on form 13f. Open-ended funds (these are what most people think of as mutual funds) are traded through dealers. The value of the fund is based on the values of the pool of securities held by the fund. ETFs and UITs have their own tradable shares and can be valued independently of their holdings. The growth of such funds as used by institutional investors has been dramatic in recent years, from a total of $917.794 billion invested in 2004 ($272.830 billion in equities) to $1.902 trillion in 2012 ($576.645 billion in equities) (ICI, 2013, p. 199). As a comparison, the total value of the US stock market is set at about $21.4 trillion (QVM Group Perspectives, 2012).

An ETF is an investment company whose shares trade on a stock exchange the same as those of any publicly traded equity ($1.3 trillion in net assets in 2012, (ICI, 2013, p. ii). The first ETF was created in 1993. It was not until 2008 though that the Securities Exchange Commission (SEC) allowed ETFs to function that were not just index funds (ICI, 2013, p. 46). An ETF may have significant diversification by holding equities, commodities, bonds and real estate.

The direct precursor to open-ended mutual funds in the United States was the Foreign and Colonial Government Trust founded in London, United Kingdom, in 1868, as an investor in government bonds. The first mutual fund in the United States (the Massachusetts Investment Trust) was established in Boston in 1924. Early funds were concentrated in real estate and public utilities. Regulation of the funds was not established until 1934, with the creation of the SEC. Open-ended funds were slow to become popular. It was not until 1951 that the number of mutual funds exceeded 100 for the first time (ICI, 2013, pp. 225–226). By 2012, mutual funds' assets exceeded $13 trillion dollars. These asset figures include both institutional and individual investor accounts (ICI, 2013, p. 198).

UITs are close-ended funds that are also traded on exchanges ($72 billion in 2012, ICI, 2013, p. ii). They differ from ETFs in that once the fund's portfolio is purchased, it is not changed until the pre-determined expiration of the fund, whereon the investment is returned. These are not to be confused with the investment trusts (also close-ended funds) started in the United States in the 1920s and were very speculative in nature and were not meant to generate long-term returns (Galbraith, 1961). Close-ended funds tend to invest in less-liquid assets. The value of close-ended funds was $265 billion in 2012. Bonds typically make up the majority of a fund's portfolio (ICI, 2013, p. 63).

Insurance companies

Insurance companies come in two types relevant to institutional investment and form 13f. Public insurance companies issue shares and are traded on exchanges like any other stock. These companies are included within 13f. Most large insurance companies are of this type. The oldest public insurance company was Insurance Company of America (1792) in Philadelphia (minyanville.com, 2013). Since the beginning of the twentieth century, investment banks and public insurance companies have had significant financial interconnections (Griesst, 2012). The banks underwrote securities that were subsequently sold to the insurance companies.

Mutual insurance companies are owned by policy holders and are not subject to inclusion in 13f reporting. America's first mutual insurance company was the Philadelphia Contributionship for the Insurance of Houses from Loss by Fire in 1792 (The Philadelphia Contributionship, 2013). The oldest public insurance company was Insurance Company of America (1792) established in Philadelphia (New York Times, 1905).

Pension funds

The first public pensions were those granted to veterans of military service from colonial times. In the late nineteenth century, pensions were extended to some state and local workers. Some larger cities followed this trend after 1850. The first state pension plan for general state employees was established in 1911 in Massachusetts (Clark, Craig, & Wilson, 2005).

Historically, American Express Company established the first US corporate pension plan in 1875, but few employees qualified (Latimer, 1932). The first large industrial pension fund was Carnegie Steel Pension in 1901 and within the next 30 years, 329 industrial firms had pension plans covering 10% of the labour force. Most of the growth of private pension plans has occurred since World War II. By 1982, about one-half of private sector workers and three-fourths of government workers were enrolled in pension plans (Investments and Income.com, 2013) to 57% of workers by the end of 2010 (Blume & Keim, 2012, p. 4). Total retirement market assets are estimated at $19.5 trillion in 2012 (ICI, 2013, p. ii).

Hedge funds

The first hedge fund was started in 1949 but saw little growth until the 1960s. Even by 1990, there were still as few as 300 funds (Hedgeco.net, 2013). Hedge funds are private investment partnerships that are open to only some investors. The investors must have at least $1 million in net worth and have significant investment knowledge. Hedge funds are illiquid since they often require investors to hold their investment for at least one year. The hedge fund's portfolio may be much more diverse in the types of its holdings than mutual funds (investopedia.com, 2013).

Sovereign wealth funds

A sovereign wealth fund "is a state-owned investment fund or entity that is commonly established from balance of payments surpluses, official currency operations, the proceeds of privatizations, governmental transfer payments, fiscal surpluses, and/or receipts resulting from resource exports" (SWF Institute, 2013). Most sovereign wealth funds do not publicly reveal their portfolio holdings. The number of sovereign wealth funds represented in the 13f filings is very small.

Past research

Corporate control

We argue that an incredible amount of power has been transferred from individuals across the United States to a few firms concentrated in select geographic locations making institutional investment an important research avenue for economic geographers. This paper examines the geography of command and control of the institutional investment industry following research long pursued by James Wheeler and other economic geographers on the location of corporate influence and decision-making power (Borchert, 1978; Goodwin, 1965; Wheeler & Mitchelson, 1989a, 1989b).

This category of research was perhaps initiated by Goodwin (1965), who highlighted the dominance of New York City as a location for corporate headquarters, but also offered conditions for other cities to realize comparable success. Research escalated in the 1970s when Gottman (1970) coined the "quaternary" concept to denote high-level corporate decision-making as a unique economic activity. The theory of headquarters evolution of Semple and Phipps (1982) followed to provide a significant leap forward in explaining the implications for the quaternary sector temporally. Their research recognized a series of regional headquarter cities in the United States, and rationalized the emergence of these cities over time. It should be noted the theory only applies to large economically developed countries. Wheeler (1990) and Wheeler and Mitchelson (1989a), Wheeler and Mitchelson (1989b) also added significant contributions to understanding this phenomenon.

Divergent categories of research have resulted from these traditional headquarters analyses. Some have focused on the spatial distribution of the actual decision-makers of the corporations, the directors (O'Hagan & Rice, 2011, 2013) and interlocking directorates (Green, 1983; Green & Semple, 1981, 2004, O'Hagan, Rice, & Green, 2008; O'Hagan & Green 2002). However, the work of Rice and Semple (1993) on the Canadian urban system is the lone study that links the network of directors' research to urban corporate control.

Perhaps the most renowned research applicable to this study surrounds World Cities research (Beaverstock, Taylor, & Smith, 1999; Knox, 1995; Sassen, 2001). Unfortunately, much of this research and those interrelated to it often study several industries collectively. This of course has its limitations. For example, O'hUallachain (1985) revealed a geographical shift in foreign direct investment from the northeastern United States to the southwest region but noted a difference between venture capital investment and that associated with manufacturing. Graves' (1998) comparison of the mutual fund industry's characteristics to other financial industries is another illustration. His findings suggest that mutual fund assets were not concentrated in the largest urban centres like the other industries. In the end, he argued for a re-evaluation of the quaternary location theory. Consequently, geographers such as Bodenman (2000) have demanded research on specific industries so that we can better comprehend the differences in location theory. This present study on institutional investment takes up their call.

Botts and Patterson (1987) perhaps provided the first research specifically highlighting the geography of institutional investment. They examined the spatial distribution of pension fund investments controlled by the State of Wisconsin. They found an increasing amount of investment was directed towards the sunbelt and southwestern regions of the United States and further away from Wisconsin and snowbelt states. As a result they called into question the investment board's policy of investing in Wisconsin, going as far as to argue this undercut the state's competitive position.

Green and Meyer (1992) provide an analysis of the location of investment companies for the period 1940–1985 for open- and close-ended mutual funds. New York City was the dominant location followed by Boston. These two cities were defined as the locational core of the companies. The rest of the 185 cities with funds headquarters were divided into three groups: semi-core, semi-periphery and periphery. It was further shown that the core and semi-core cities had increased their representation since 1940.

Green, Meyer, and Cromley (1992) followed this up with a study of the Canadian institutional investment industry for 1990. The Canadian system unlike the US system was dominated by public pension plans. Toronto was the dominant location of funds followed by Montreal. Most investment was intra-country at 62%, followed by the US investment at 22.5%.

A study of institutional stock ownership in the United States in 1980 shows a hierarchical urban system dominated by firms located in New York, Chicago and San Francisco (Green, 1993). Green used a network blocking model developed by Everett (1983) to define six classes of urban centres. Cities were classified by the strength of their investment between cities. The three most important cities were New York, Chicago and Boston in that order.

Green (1995) provides an update of his previous analyses with institutional investment data for 1990. Using a multiple classification analysis of the dollar values of linkages between cities and type of investment firm, New York City is dominant. Boston is the second most important, followed by Los Angeles and San Francisco. Unlike other studies, Chicago is ranked seventh, but this may be due to the data sampling instead of population data. Public pension plans on average had the highest valued portfolios.

Bodenman (1998, 2000) examines investment advisory firms for tax-exempt fund sponsors (essentially pension plans). He states that such firms are not necessarily tied to the financial core of large metropolitan areas. Some of this is because state public plans are locationally tied to their respective state capitals. New York City has experienced a decline in its percentage of advisory firms of 28% in 1983 to 18.7% in 1996 (p. 44). This is consistent with the findings of Gong and Keenan (2012) showing that Manhattan has lost employment in financial services.

Death of geography in finance

The current structure of the global financial system often results in declarations of a diminishing role for geography in explaining its form. This assertion is based on significant improvements in communications and the resulting increase in capital mobility. So pervasive are these changes that researchers like O'Brien (1992) claimed a death for geography. Moriset and Malecki (2009) add that technological improvements and the evolution of the information society challenge the agglomeration theory, so fundamental to much of economic geographers' research. It is a fundamental assumption of this argument that information is ubiquitous. In other words, the concept of place does not matter for institutional investors, because no investor can profit by virtue of their particular location.

Research has shown that institutional investors are remarkably concentrated in core cities of the United Kingdom (Martin & Minns, 1995), Australia (Ramsay & Blair, 1993; Torrance, 2009a, 2009b) and Canada (Green, et al., 1992; Figure 1 this paper). In the United States, perhaps the industry is more decentralized, but previous research still shows that it is concentrated in a few cities such as New York, Boston and Chicago (Green, 1993, 1995; Green & Meyer, 1992). These same institutional investment firms also concentrate their assets in domestic firms. The study by French and Poterba (1991) is probably the most prominent among many studies documenting the home country bias. They found that 94% of US equity traders' money was allocated to domestic securities, labelling it the home bias puzzle. Coval and Moskowitz (2001, 1999, p. 2046) explain this phenomenon by suggesting "when capital crosses political and monetary boundaries, it faces exchange rate fluctuation, variation in regulation, culture, and taxation, and sovereign risk, which many home bias explanations focus on as the primary factors discouraging investment abroad". Even for

Figure 1. Institutional investment by city of origin, 2010.

international transactions, research has shown that geography is a contributing explanation for these transactions (Portes & Rey, 2005).

With respect to portfolio investment, it is well documented that individuals hold a geographical bias towards their holdings. Huberman (2001) suggests that familiarity of the local business climate breeds investment. Heath and Tversky (1991) call this the "competence hypothesis", and suggest that investors are sometimes willing to forego the advantage of diversification and concentrate on a few companies with which they are presumably familiar. Clark (2010, 2011) rationalizes this by remarking that people are simply poor at searching for relevant information and tend to act based on immediate experience. He calls people myopic, and therefore the local environment is too influential. A recent study shows that 80% of companies prefer to locate investment managers close to their investment target (Piggott, 2012).

Much of the previous research that includes a spatial component is devoted to individual investors who are generally less knowledgeable about global opportunities. Interestingly, research has shown that this spatial pattern holds even for more knowledgeable institutional investors. This of course begs the question, why? Investors may prefer geographically proximate investments arising from several potential sources. For instance, investors may feel more comfortable about local companies or firms they hear about a lot or they may have a psychological desire to invest in the local community. Traditional economic models, though, suggest that a superior explanation is the fact that human decision-making is related to maximizing the expected payoff and minimizing risk (Tversky & Kahneman, 1974).

Investors' aversion to risk suggests that these portfolios should be diversified, particularly internationally. Nonetheless, they greatly ignore this advice—by and large investors' money stays in their home countries and goes to geographically close firms. One can see why, as Choe, Kho, and Stulz (2005), Dvořák (2005), Teo (2009) and Stotz (2011) all find that international funds perform more poorly than domestic funds.

The paramount answer lies in the fact that maximizing profit or decreasing risk requires the generation, manipulation, interpretation and flow of information. Consequently, the acquisition of market information is imperative. This is where geographical theories can help explain a concentration in the industry. Geography matters because contrary to O'Brien's position, information, or at least information that leads to competitive advantages, is not ubiquitous. It embodies key factors in the production process of financial products at the regional level, even if these products can be processed and distributed worldwide more easily than ever before. In perhaps the most extensive study, Bae et al. (2008) use a 32-country sample to find that resident analysts in a country make more precise earnings forecasts for firms in that country than non-resident analysts. These superior forecasts are attributable to superior information.

It has been shown time and again that some types of information are spatially sticky (Markuen, 1996). This follows Morgan (2004), who argues that the "death of geography" is exaggerated and that information transfer is dependent upon proximity. If space is irrelevant, then why is it important for companies to locate in specific cities? This discussion dates from Marshall (1920) and Hoover (1948), researchers who were among the first to establish that firms can benefit from knowledge spillovers. Once a critical threshold is passed, many new firms will opt for this region to take advantage of agglomeration economies. For institutional investment, local investors access knowledge by talking to employees, managers and suppliers of firms. They may obtain important information from the local media. They may have close personal ties with local executives (Coval & Moskowitz, 1999, p. 2046). Returning to the purpose of this study, if information is so important to the industry, does it not make sense for it to remain highly concentrated so that institutional investors can take advantage of information spillovers, which are more likely at smaller geographic scales?

Informal networks involve trading information between professionals in firms that may or may not be rivals (Schrader, 1991; von Hippel, 1987). Information trading is an informal practice because it occurs among employees acting on their own, rather than as a corporate strategy. Teo (2009) found that funds with a physical presence (headquarters or research office) in their investment region outperform others. The local information advantage is pervasive across all major geographical regions, but is the strongest for emerging market funds and funds holding illiquid securities. As Agnes (2000, p. 367) notes:

> Trust is important, especially on the point of discretion. You need to know that you can trust brokers and that if you need to have discretion in executing an order, they will work the order to your advantage. You need to know that you can trust them with your order and trust their judgment also. You want information flows about who's doing what, and how aggressively. And there are other people that I would never give that order to. The brokers that I deal with now, I've known for four years. (Interview with swaps dealer, 14 February 1995),

> Traders know each other, but they can't talk too much—it's all confidential. You can't talk about the latest trade, or the new product. Nonetheless, they're all pretty buddy-buddy. They're always talking to each other. If we want to do a swap with [Rival Bank A], our trader knows the guys there and might persuade them to help us. (Interview with associate director, Debt and Capital Markets, 2 May 1995.)

These networks are underpinned by social relationships, which require face-to-face interaction and have not been replaced by advances in technology. Contacts are made through venues where face-to-face contact provides opportunities to make judgments about the quality of information and expertise of potential collaborators in solving specific contextual problems (Fleck, 1996).

Formal dealing networks between financial firms are circumscribed by formalized institutional practices defined by market conventions. Formal networks are embedded in informal relationships among individual financial professionals who facilitate transactions and diffusion of market information (Thrift, 1994; Thrift & Leyshon, 1994). These relationships are increasing in importance in a global world, not decreasing.

Work by Coval and Moskowitz (2001) shows US mutual fund managers are better at picking stocks of nearby firms than stocks of distant firms. Ivković and Weisbenner (2005) witness a similar pattern with US retail investors. However, Malloy (2005) finds that US equity analysts issue more accurate earnings forecasts for nearby firms than for distant firms.

Academics and practitioners have long been interested in understanding institutional investment firms' information advantages in stock investments and the impact of such advantages on stock returns. Yet, although a growing body of literature has examined these issues extensively, the results so far are inconclusive. For example, while several studies show that certain groups of institutional investors, such as mutual fund managers, can capitalize on their superior information (Baker, Litov, Wachter, & Wurgler, 2010; Daniel, Grinblatt, Titman, & Wermers, 1997; Grinblatt & Titman, 1989, 1993), another line of research finds that mutual fund managers underperform appropriate risk-adjusted benchmarks (Jensen, 1968; Malkiel, 1995; Gruber, 1996; Carhart, 1997). Mainstream research on institutional investors has mostly taken a functionalist approach, by which the main research agendas of this classify institutional investors, study their liabilities, investment techniques and performance in given financial institutions (Davis & Steil, 2001).

Data

The data for this research come from Form 13(f), which is information submitted to the Securities and Exchange Commission. By law, an institutional investment manager that uses the US mail (or other means or instrumentality of interstate commerce) in its business and exercises investment discretion of $100 million or more in Section 13(f) securities must report its holdings (SEC, 2012a). The enabling legislation specifies that the SEC may impose a $10 million reporting limit if it so desires. Currently, it has set the reporting limit at $100 million. If an institutional investor exceeds the $100 million limit at any point during a given year, it must report holdings for all four quarters. Therefore, some smaller investors who do not normally report may be occasionally caught by the requirements. This will force the disclosure of smaller investments not normally added to such data, filling out the lower end of the investment values distribution. Section 13(f) securities are those registered on a national stock exchange (Library of Congress, n.d.; SEC, 2012b). Section 13f of the Securities Exchange Act (SEA) was enacted by Congress in 1975. The history of Section 13f of SEA is the result of many legislative attempts to reduce the perception of corruption and collusion in government and business by increasing transparency, especially in the wake of the then ongoing Watergate scandal (Shaw, 1981).

An examination of the data reveals that some institutional investors with less than the $100 million threshold also voluntarily file, thereby providing some coverage of the lower end of the industry. This is an important restriction of the data in that only publicly traded securities are reported and that other investments by the institutional investors need not be revealed. Estimates of portfolio size are therefore by nature conservative in value.

On 21 December 2010, there were 15,513 securities in the list of 13(f) securities. The data were parsed from filings held by the SEC by www.whalewisdom.com (2011). Stocks trading for less than $0.50 per share were deemed to be penny stocks and were removed from the sample, since serious institutional investors would not participate in such trading. Firms declared to be bankrupt or no longer traded on any major exchanges are excluded, as are index funds (since investment decisions are dictated by the composition of the index) and pink sheet stocks (stocks not traded on an exchange and do not file with the SEC). Mutually held insurance companies and mutual funds are exempt from reporting, so they are also excluded.

For each institutional investment, the 13(f) forms provide information on the name and headquarters address of the institutional investment firm (including country), the name of the investee company, its headquarters address, its industry, the number of shares held by the investor, the declared value the stock, the investee ticker symbol and the date of the filing. Earnings per share was added to the data by using the ticker symbol to match financial data from Capital IQ Compustat. Unfortunately, 13(f) does not classify the investors as to type.

Analysis of firms

This paper contributes to the literature that explores the relationship between distance and investment performance by examining the US institutional investment industry at two different levels of aggregation. The individual institutional investment firms are explored before focusing on their geographic distribution of headquarters. Table 1 provides a list of the 26 largest institutional investors in the United States, two Canadian, one Swiss and one German-based filing firms covered by Section 13(f) requirements. These investors account for more than 50% of the total reported investments. The world's largest investor is the Blackrock Group with over $915 billion (over 12.2% of the total) in declared

Table 1. Largest institutional investors in the US equities, fourth quarter, 2010.

	Firm	Value of investments (billion $)	City
1	Blackrock Group of Companies	915	
	Blackrock Institutional Trust Company N.A.	*383*	*San Francisco*
	Blackrock Fund Advisors	*195*	*San Francisco*
	Blackrock Group Ltd	*120*	*London*
	Blackrock Advisors LLC	*87*	*Philadelphia*
	Blackrock Investment Management LLC	*58*	*Philadelphia*
	Blackrock Inc.	*44*	*New York*
	Blackrock Japan Co. Ltd	*28*	*Philadelphia*
2	Vanguard Group Inc	583	Philadelphia
3	State Street Corp	573	Boston
4	FMR LLC	510	Boston
5	Bank of New York Mellon Corp	442	New York
6	Price T Rowe Associates Inc /Md/	288	Washington
7	Capital World Investors	283	Los Angeles
8	Wellington Management Co Llp	265	Boston
9	Capital Research Global Investors	247	Los Angeles
10	Morgan Stanley	245	New York
11	J P Morgan Chase & Co	185	New York
12	Invesco Ltd.	172	London
13	Franklin Resources Inc	137	New York
14	Alliancebernstein L.p.	132	New York
15	Ameriprise Financial Inc	131	Minneapolis
16	Goldman Sachs Group Inc	128	New York
17	Wells Fargo & Co/Mn	124	San Francisco
18	TIAA Cref Investment Management LLC	111	New York
19	Bank of America Corp /De/	101	Charlotte
20	Janus Capital Management LLC	101	New York
21	Deutsche Bank Ag\	93	Frankfurt
22	Royal Bank of Canada \	92	Toronto
23	Massachusetts Financial Services Co /Ma/	88	Boston
24	Dodge & Cox	83	San Francisco
25	Dimensional Fund Advisors Lp	78	Austin
26	Jennison Associates LLC	73	New York
27	Geode Capital Management LLC	72	Boston
28	UBS Ag	70	Basel
29	Neuberger Berman Group LLC	69	New York
30	TD Asset Management Inc	68	Toronto
	Total	6,459	
	Other	1,018	

investments. The Blackrock Group's parent company Blackrock Inc. is found in the premier institutional investment city of New York with subsidiaries located in the other top investment centres such as London, Philadelphia and San Francisco. As the table demonstrates, the distribution of firms by value of investment declines rapidly. The second largest company, Vanguard Group, is about 64% of the value of Blackrock Group. Institutional investment is highly concentrated in a small fraction of the total number of reporting investment firms. This of course could also have implications for spatial concentration of the industry, as discussed later.

Several foreign-based firms are active in the US market: Deutsche Bank AG (eighth in the world (Germany), the Royal Bank of Canada (Canada's largest, 38th in the world), UBS AG (20th in the world, Switzerland) (Global Finance, 2012) and TD Asset

Management (parent corporation, TD Bank, 48th in the world, Canada) (Global Finance, 2012). Most of the top 50 banks have representation in the US market while only a single pension plan is represented in the top 30, TIAA-CREF with $111 billion. Twenty-four other North American public pension plans have substantial presences in institutional investments with the total value of investments being $516.4 billion. Canadian plans accounted for about $34.9 billion. This is because the Canadian capital markets are too small to absorb large investments. Therefore, Canadian institutional investors must go outside of the country.

The top 20 target stocks are found in Table 2. Not surprisingly Apple, which has been the darling of institutional investors, heads the list. All of the 20 are members of both the Fortune 500 and the S&P 500 with all major industrial sectors represented. The number of institutional investment firms with holdings in the top 20 stocks is 1,540, almost one-half of the total of all of investors. The smaller level of the larger institutional investors shows that institutional investment has a more limited effect on changing the status quo of the existing universe of investee firms than might have been expected looking at the institutional investment industry as a whole.

Examination of the largest holders shows that State Street Corp. (ranked third on Table 1) is the largest holder of stock 50% of the time. This shows a willingness to hold larger volumes of shares than is normal for most institutional investment firms. The other major firms from Table 1 are FMR (rank 4), Capital World Investors (rank 7), Capital Research Global Management (rank 9) and Morgan Stanley (rank 10). The pattern is not to invest heavily in any one firm, this being consistent with a diversified portfolio. The

Table 2. Top firms held by institutional investors, 31 December 2010.

Rank	Company	Institutional investment (billion $)	Number of investors	Largest holder of common shares	% of investor portfolio	% held by investors
1	Apple	208.7	1,649	FMR	3.43	77
2	EXXONMobil	176.3	1,732	State Street	2.51	49
3	Microsoft	145.9	1,792	Capital Research Global Mgt	3.45	64
4	Wells Fargo and Co	129.8	1,316	Berkshire Hathaway	20.2	77
5	JP Morgan Chase	121.9	1,565	State Street	1.17	74
6	Google	119.4	1,331	FMR	1.92	81
7	Chevron	113.8	1,524	State Street	1.62	63
8	IBM	107.1	1,579	State Street	1.63	61
9	Johnson and Johnson	105.0	3,270	State Street	1.48	64
10	Procter and Gamble	102.5	1,593	State Street	1.23	58
11	General Electric	98.6	1,325	State Street	1.24	52
12	Oracle	96.3	1,531	Capital Research Global Mgt	2.94	62
13	Pfizer	96.2	1,384	State Street	1.17	71
14	Coca Cola	96.2	1,378	Berkshire Hathaway	25.03	56
15	AT&T	92.1	1,266	Cullinan Associates	1.03	58
16	Berkshire Hathaway A and B shares	91.8	1,292	Stewart West Indies Trading	99.92	48 A shares only
17	Schlumberger	86.5	1,383	State Street	0.78	79
18	Merck and Co	80.5	1,027	Capital World Investors	2.67	74
19	Bank of America	79.8	1,309	State Street	1.06	60
20	SPDR S&P 500 (ETF)	75.3	1,568	Morgan Stanley	5.81	na

two major exceptions being Berkshire Hathaway's heavy exposure to Coca Cola, consistent with Warren Buffett's strategy of making large investments in firms (Forbes, 2012) and the almost 100% investiture in Berkshire Hathaway by Stewart West Indies Trading Company (116th largest investor).

Analysis by city

The core metropolitan statistical areas (MSAs) definition by the Bureau of Management and Budget was used to define the membership of cities and towns within the largest 10 cities. Cities below the top 10 were defined by MSAs (Nussle, 2008). Any headquarters locations below the definition of MSA were defined as rural and were not included in the urban analysis.

Figure 1 shows the distribution of the value of institutional investment by urban area that clearly reveals a concentration of investment in a few of the largest cities. The coasts and the Eastern Mid-west are well represented. Surprisingly, the growth areas of the southeast and southwest possess significantly less, which suggests that institutional investment firms exhibit substantial levels of spatial inertia. Such inertia is fuelled by their long history in their current locations and the informational network that implies.

Table 3 provides a listing of the 10 cities housing the most headquarters of institutional investors. As is well known, New York City is the dominant financial centre of the United States, and so it is for institutional investment. The city's prominence came early, for by 1785, New York City was already the location of the offices of large merchants, the three largest banks and the three largest insurance companies (Depew, 1895, p. 56). Within New York City, offices are concentrated in Mid-town Manhattan, particularly just south of Central Park and the Wall Street district at the southern tip of the island (Figure 2). The firms in Manhattan account for 75%–80% of the investment firms in New York City.

Figure 2 visually underrepresents the concentration, particularly in Mid-town, because there are many instances of investment firms sharing the same building but on different floors. Both the Mid-town and Wall Street District concentrations are thought to be the consequence of residential and manufacturing patterns coupled with public transportation hubs and not the quality of the underlying bedrock (Barr, Tassier, & Trendafilov, 2011).

The Wall Street concentration is also the legacy of the period when financial trading required a physical presence at the New York Stock Exchange (NYSE) as well as being

Table 3. Top 10 origin cites for institutional investment, 2010.

City	Investment value (billion $)	Number of links	Percentage of total investment all cities	Population rank (United States and Canada)*
New York	2,365	153,532	18.6	1
Boston	1,930	51,730	14.8	11
San Francisco	950	30,074	7.3	12
Foreign	932	62,814	7.2	n.a.
Philadelphia	826	35,161	6.4	5
Los Angeles	823	16,752	6.3	2
Chicago	603	43,429	4.6	3
Washington, DC	362	11,945	2.8	7
Toronto	347	17,454	2.7	9
Atlanta	249	9,151	1.9	10
Total	9,391	432,042	72.4	

*US Census Bureau (2012), Statistics Canada (2011).

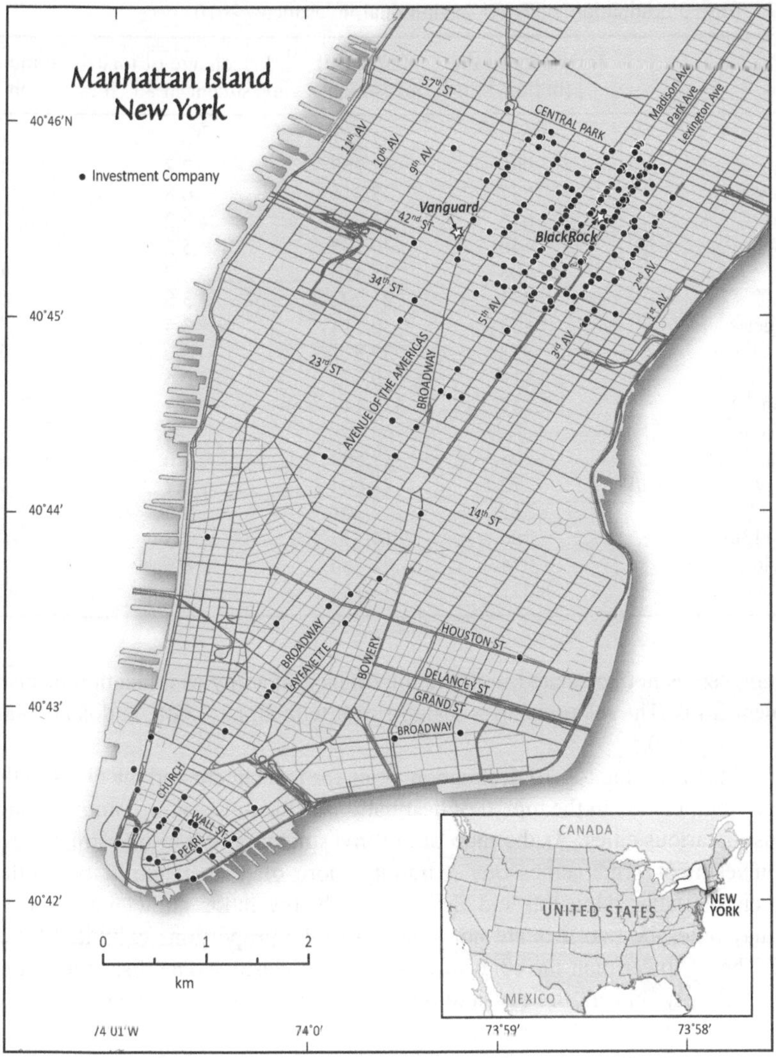

Figure 2. Institutional investment firms in Manhattan, 2010.

the location of the most important bank, J.P. Morgan (now J.P. Morgan Chase), for much of the United States' financial history. As Geisst (1997) noted, the historical geography of the banking industry provided New York City with an early dominance of banking. This concentration was strengthened in 1929 when the NYSE allowed the listing of investment companies for the first time (Galbraith, 1961).

Mid-town Manhattan's attractiveness was enhanced in the late 1980s with the upgrades on the area's telecommunication systems to a fibre optic network. This was further enhanced with the construction of the trans-Atlantic TAT-8 fibre optic cable by AT&T in 1988 (Bank Administration, 1989) and several subsequent, newer and higher-capacity cables.

The cities of Boston, Chicago and San Francisco show similar clustering in their down towns. Importantly, membership in this top 10 is not entirely related to population size. Miami (eighth largest) and Dallas-Ft. Worth (fourth largest) are not included. Toronto, Canada's largest city, is included, as is San Francisco (12th). Results reveal that

Table 4. Top 10 destination cities for institutional investment, 2010.

City	Investment value (billion $)	Number of investments	Percentage of total investment in all cities	Ratio of out to in investment
New York	2,067	92,391	16.7	1.20
Foreign	968	62,257	7.8	1.00
San Francisco	837	38,316	6.8	1.20
Houston	646	34,210	5.2	0.06
San Jose	646	20,505	5.2	0.06
Chicago	578	33,413	4.7	1.10
Dallas	471	21,292	3.8	0.21
Los Angeles	433	30,114	3.5	2.00
Boston	372	29,877	3.0	5.40
Seattle	319	10,772	2.6	0.20
Total top 10	7,337	373,147	59.3	
Below 10th				
Philadelphia (15th)	203	17,504	1.6	4.00
Washington, DC (14th)	272	21,604	2.2	1.30
Toronto (11th)	313	7,301	2.5	1.10
Atlanta (13th)	286	14,777	2.3	0.83
Total below 10th	1,074	61,186	8.6	
Grand total	8,411	434,333	67.9	

population size is not correlated with the importance of a city as a location of institutional investment firms. The relationship is insignificant with an $r_p = .048$, and is not statistically significant at $p \leq 0.5$.

Table 4 provides the top 10 destination cities for investment and the cities in the top 10 of origins not included in the top 10 of destinations. This reflects the command and control positions of various cities. As the ratio of out-investment versus in-investment reveals, the largest investment firms' cities buy extra-city more often than intra-city, while growth areas such as Dallas, Houston and San Jose with low ratios attract investment. As the reader may note, the two lists are not identical to the proportions exhibited by each city generally being lower than the concentration of the investors. The distribution of destination cities closely mirrors the distribution of available equities. The coefficient of determination between the market value in equities invested in each city and the total available market value of stocks in those same cities is $r^2 = 0.98$ ($n = 242$, out of a possible 371 cities) with reported investments equalling more than 75% of the total available (6,853 stocks with total market value reported). The high r^2 is the result of the dominance of institutional investment in the market. Such results suggest a very efficient market from a spatial perspective or constraints imposed by the large size of the institutional investment firms who do not want to load too heavily into any single firm.

Although the pattern of investment is spatially uneventful, the pattern of the rates of returns on equity is much more uneven. Figure 3 provides a map of the positive rates of return in equity investments made in the target cities. Rates of return were calculated by multiplying the earning per share (calculated from number of shares held/reported value) by the number of shares held in each investment to provide a weighted sum and then dividing by the number of total shares held within the metropolitan area.

Only 12 cities possessed firms that produced negative returns. Table 5 reveals that 11 of these cities are considered small, with San Jose being the sole exception. Table 5 also lists the cities with the best rates of return. Again they are smaller centres except Indianapolis. The lack of the largest cities in either list represents a regression to the

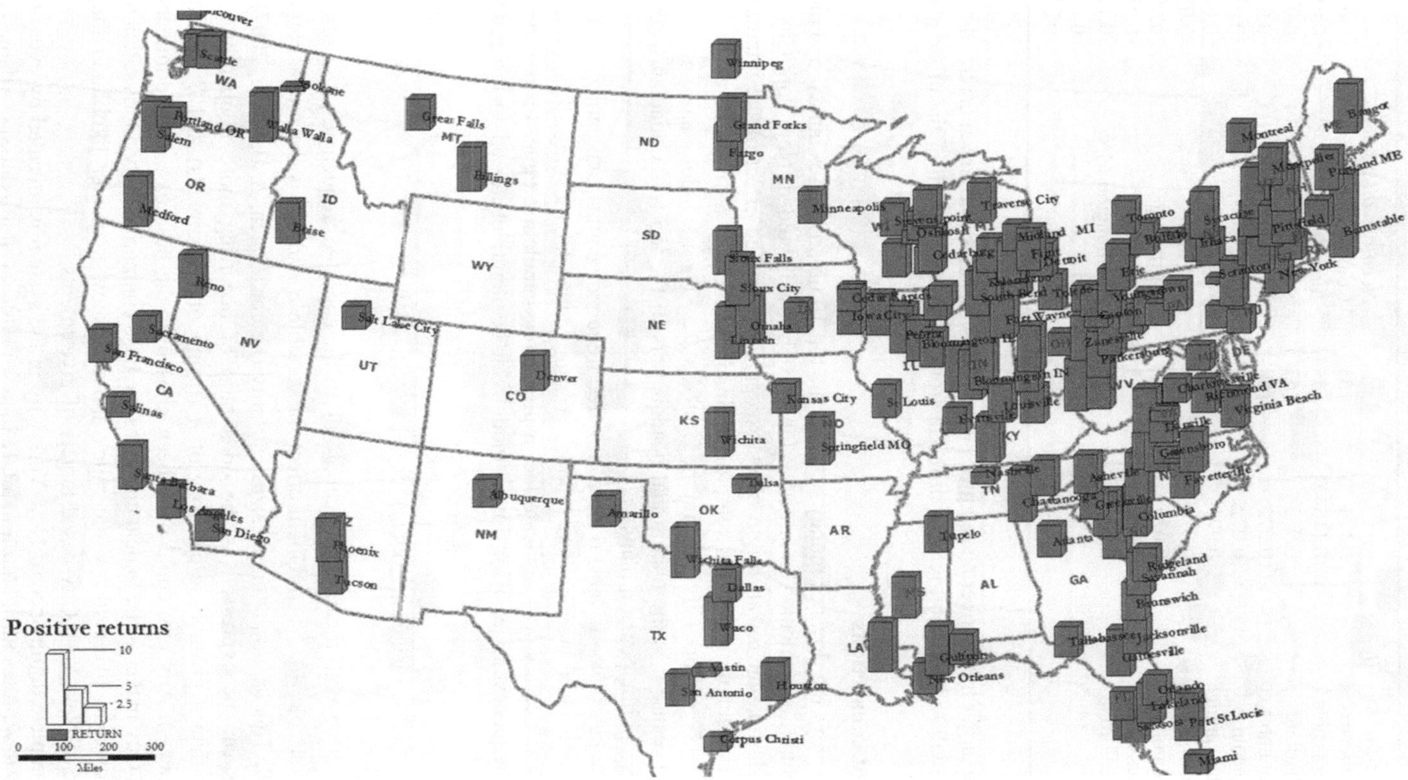

Figure 3. Positive returns on investment by target city, 2010.

Table 5. Rates of return by target city, 2010.

	City	Rate of return	Rank	City	Rate of return
	Positive top 10			*Negative, all cities*	
1	Indianapolis	0.124	165	Colorado Springs	−0.001
2	Barnstable	0.115	155	Juneau	−0.013
3	Charlotte	0.072	167	Santa Fe	−0.028
4	Columbia	0.072	168	San Jose	−0.030
5	Lincoln	0.072	169	Naples	−0.033
6	Chattanooga	0.071	170	Little Rock	−0.049
7	Hutchinson	0.071	171	Jackson	−0.058
8	Bangor	0.070	172	Harrisburg	−0.065
9	Glens Falls	0.070	173	Memphis	−0.087
10	Medford	0.069	174	Bloomington MN	−0.172
			175	Bend	−0.209
			176	Elkhart	−0.210

mean. The larger the number and values of portfolios hosted by an urban centre, the more difficult it will to be exhibit returns different from the overall return of the urban system.

Interurban connections

Table 6 displays the flows of money between the top 10 urban centres ranked by holdings of institutional investment. The largest value in each row is bolded. New York City is the dominant destination for all origin cities except Toronto. The effect of the international border is evident with Toronto's more dominant position as a financial centre within Canada, than New York City's within the United States. Toronto's dominance in Canadian finance is because it is the operational headquarters of the top five Canadian banks and several large public pension plans.

If the money flows of Table 6 are considered frequency counts, standardizing the table is possible, via a saturated log linear model. A saturated model, meaning all four terms are included, was necessary to replicate the original cross-tabulation table. In its form for a two-way frequency table, a log linear model can be iteratively fitted to the following equation:

$$\ln(f_{ij}) = \mu + \lambda_i^A + \lambda_j^B + \lambda_{ij}^{BA}$$

where μ is the overall mean of the natural log of the expected frequencies, λ_i^A is the main effect for variable A, λ_j^B is the main effect for variable B and λ_{ij}^{AB} is the interaction effect of A and B. This technique emphasizes causality (Pearce, Green, & Baldwin, 2007). The parameters can be expressed as the odds ratio of occurring more frequently or less frequently than a given base category of A (origin of investments) and B (destination of investments). Any city could have been chosen as the base case. The dominance of New York City makes it the logical choice as the base comparison city. Odds ratios of less than 1 show less likelihood, while ratios above 1 show a greater likelihood than the occurrence of New York City-based targets of investments.[1]

The complete dominance of New York City in the system is revealed by every origin and destination odds ratio being much less than 1 (Table 7). For example, Atlanta is only as 10% likely to be an origin of institutional investment than is New York City. The lone exception is Boston, verifying the city's current and historical importance in the United

Table 6. Flows between top 10 metropolitan areas in billion $, 2010.

Origin/destination	Boston	San Francisco	Philadelphia	Los Angeles	Chicago	DC	Toronto	Atlanta	New York City	Total
Boston	6.7	14.5	2.9	7.1	8.3	4.5	2.2	4.4	**33.1**	83.8
San Francisco	2.6	6.5	1.5	3.2	3.9	1.9	1.2	2.0	**15.3**	38.2
Foreign	2.2	5.7	1.1	2.5	3.6	1.5	1.9	1.9	**15.1**	35.7
Philadelphia	2.5	5.9	1.6	3.1	4.2	2.1	0.2	2.1	**14.9**	36.6
Los Angeles	2.5	6.1	1.0	2.7	4.0	1.3	1.2	2.2	**13.8**	34.9
Chicago	1.7	4.0	1.0	1.9	3.3	1.4	2.0	1.2	**9.2**	25.6
Washington, DC	0.9	2.9	0.4	1.1	1.4	1.4	0.2	0.8	**5.4**	14.4
Toronto	0.3	0.9	0.2	0.4	0.6	0.3	**10.1**	0.3	2.5	15.6
Atlanta	0.8	1.5	0.5	1.0	1.1	0.5	0.4	0.9	**3.5**	10.2
New York	7.1	13.4	3.9	7.3	10.8	4.6	3.4	4.3	**35.1**	89.9
Total	27.4	61.4	14.1	30.4	41.2	19.6	22.8	20.1	147.9	384.9

Note: Bold values represent the largest row value.

Table 7. Log-linear estimates for primary effects (saturated model).

Effect	Parameter	Estimate	Odds ratio
Origin	Boston	−0.06	0.94
	San Francisco	−0.83	0.44
	Foreign	−0.84	0.43
	Philadelphia	−0.85	0.43
	Los Angeles	−0.93	0.39
	Chicago	−1.34	0.26
	Washington, DC	−1.87	0.15
	Toronto	−2.65	0.07
	Atlanta	−2.31	0.10
	New York	0.00	1.00
Destination	Boston	−1.59	0.20
	San Francisco	−0.96	0.38
	Philadelphia	−2.21	0.11
	Los Angeles	−1.57	0.21
	Chicago	−1.18	0.31
	Washington, DC	−2.04	0.13
	Toronto	−2.33	0.10
	Atlanta	−2.09	0.12
	New York	0.00	1.00
Constant		10.47	

States' financial industry. Boston was an early competitor of New York City as a major trading centre for equities during the nineteenth century when the New York exchanges were dominated by bond offerings (Geisst, 1997). Conceptually, the origin and destination odds ratios are similar to the origin and destination effects in a gravity model.

Table 8 builds upon this premise to reveal the interaction effects as expressed by the odds ratios between city pairs that can be interpreted as the distance effect within a gravity model. However, in Table 8 the distance effect is disaggregated for individual city pairs (Willekens, 1983). When disaggregated in this way, the dominance of New York City is not present for all sets of flows. Since New York connections are the comparison cases for all links, they are set to a value of one for the odds ratio. In particular, Toronto-based firms invest in other Toronto-based firms almost 42 times more often than they invest in New York City-based investments. Chicago, San Francisco, Washington, DC, and Atlanta-based institutional investment firms all prefer local investments to those of other cities when the size effects of New York City are controlled. Localness, therefore, seems important for institutional investors in some cities. New York City of course is still the preferred location when just frequency counts of investments are the only consideration.

Conclusions

The purpose of this paper was to investigate whether geography matters for institutional investment at a time when forces have significantly internationalized operations in many industries. After initially exploring individual investors, the study focused on the geographic distribution of that institutional investment. Specifically, the study set out to learn whether this industry reveals a decentralization of operations or is it highly concentrated with a fair degree of inertia?

Building on previous research, this study revealed an underlying geography is pervasive to institutional investment. This research provides additional support for the

Table 8. Log-linear estimates for interaction effects (saturated model).

Origin/destination	Atlanta	Boston	Chicago	Los Angeles	New York City	Philadelphia	San Franciso	Toronto	Washington, DC
Atlanta	**2.18**	1.10	1.01	1.33	1.00	1.34	1.14	1.21	1.15
Boston	1.08	1.00	0.81	1.04	1.00	0.81	**1.15**	0.68	1.05
Chicago	1.04	0.91	**1.17**	1.00	1.00	0.94	1.14	2.23	1.19
Los Angeles	**1.29**	0.88	0.94	0.96	1.00	0.69	1.15	0.87	0.75
New York	1.00	1.00	1.00	1.00	1.00	1.00	1.00	1.00	1.00
Philadelphia	**1.12**	0.82	0.92	1.00	1.00	0.96	1.03	0.16	1.09
San Francisco	1.06	0.83	0.83	1.01	1.00	0.87	**1.12**	0.80	0.96
Toronto	1.06	0.62	0.78	0.84	1.00	0.58	0.97	**41.89**	0.78
Washington, DC	1.16	0.86	0.83	0.96	1.00	0.72	1.39	0.40	**1.95**

Note: Bold values represent not significantly different than 1 at $p = .01$.

importance of space in investments, particularly at the urban-area level of data disaggregation. This is particularly relevant as geography comes under increasing scrutiny associated with globalization and its forces on economic activity. At a purely descriptive level, the dominance of New York City (particularly Manhattan) overwhelms the pattern of urban-based investment. However, when the effects of this dominance are controlled, other cities present themselves as important. This is particularly so when the interurban linkages of institutional investment are examined.

A comparison with comparable studies of earlier time periods, 1980 and 1990, apparently shows that the institutional investment industry has experienced changes in three ways. The least dramatic changes are those of the urban system defined by investments. New York remains the dominant location over time. Some reordering of the other large cities is evident, but overall the urban system has been stable over the last three decades.

A great change occurred in the membership of the largest institutional investors. Of the top 10 in 1980, only one, Bankers Trust, was in the top 10 in 1990. By 2010, even it had dropped from the list. The largest investor firm in 2010, Blackrock Group, was not established until 1988. The creation of new firms in the same locales as previous firms helps maintain the geographical structure.

The last factor of change is the types of firms that participate in institutional investment. Hedge funds, demutualized insurance companies and public pension plans have increased in the portion of investment owned. Investment banks no longer completely dominate the industry, as they did in the beginning of the institutional investment industry. Firms that serve primarily as investment advisors now feature in the list of the largest investment firms.

Limitations

This study does suggest areas of possible future research. First, this study only looks at the data of the fourth quarter of 2010, and is therefore a snapshot into the early days of the stock rally in response to the Great Financial Crisis of 2008–2009. However, the vast majority of investments in the data set are positions long held by institutional investors, and are temporally dependent on previous and follow-up observations. Time-series analysis could provide deeper insights into investment patterns in all parts of the business cycle. This research has dealt with the investments after a major market downturn. Analysis of a more extended buoyant period is necessary as well.

Second, as this paper has shown, investing does have spatial attributes and previous literature such as Coval and Moskowitz (1999, 2001) show that local investors are able to attain superior returns by using local knowledge that is unavailable to outside investors, especially in small and remote locations. However, these studies only measured the linear distance between firms and investors, thus ignoring the real friction imposed by commuting between points. This study is temporally limited to the first decade of the Internet Age, and thus does not cover more recent social networking phenomena such as Facebook, Twitter and Linked-in. It would be interesting to determine whether these processes diminish the role of location in gaining superior returns on investment.

Another fruitful approach would be to disaggregate the investment data by type of institutional investment firm (i.e. pension plans vs. hedge funds) and by the industry of the held firm. It is quite likely that the investment patterns of the different kinds of investment firms differ. The different investment objectives of the classes of investment firms would almost certainly cause differences in the types of firms' stocks acquired. These differences would be evident both in an industry analysis and as having different locational preferences.

Lastly, many people who have previously worked on Wall Street such as Lewis (2010) have remarked that the investment culture of the 1990s to the present is markedly different from the investment culture that preceded it in the 1980s. In fact, they argue that the investment culture has shifted from making long-term investment based in fundamentals such as price–earning ratios and moved towards price arbitrage. Therefore, it would be appropriate to ask whether other methods of evaluating investments such as risk-adjusted returns would offer a better method of calculating local advantage rather than rate of return.

Disclosure statement

No potential conflict of interest was reported by the authors.

Note

1. A detailed description of the model calibration and theoretical basis may be found in Knoke and Burke (1980).

References

Agnes, P. (2000). The "end of geography" in financial services? Local embeddedness and territorialization in the interest rate swaps industry. *Economic Geography, 76*, 347–366.

Bae, K., Stulz, R., & Tan, H. (2008). Do local analysts know more? A cross-country study of the performance of local analysts and foreign analysts. *Journal of Financial Economics, 88*, 581–606.

Baker, M., Litov, L., Wachter, J., & Wurgler, J. (2010). Can mutual fund managers pick stocks? Evidence from their trades prior to earnings announcements. *Journal of Financial and Quantitative Analysis, 45*, 1111–1131.

Bank Administration. (1989). *A worldwide network* (Vol. 65, p. 34).

Barr, J., Tassier, T., & Trendafilov, R. (2011). Depth to bedrock and the formation of the Manhattan skyline, 1890–1915. *The Journal of Economic History, 71*, 1060–1077.

Beaverstock, J., Taylor, P., & Smith, R. (1999). A roster of world cities. *Cities, 16*, 445–458.

Blume, M., & Keim, D. (2012). Institutional investors and stock market liquidity: Trends and relationships. Retrieved from http://www.wharton.upenn.edu/jacobslevycenter/files/14.12.Keim.pdf

Bodenman, J. (1998). The suburbanization of the institutional investment advisory industry. Metropolitan Philadelphia. 1983–1993. *The Professional Geographer, 50*, 112–126.

Bodenman, J. (2000). Firm characteristics and location: The case of the institutional investment advisory industry in the United States, 1983–1996. *Papers in Regional Science, 79*, 33–56.

Borchert, J. (1978). Major control points in American economic geography. *Annals of the Association of American Geographers, 68*, 214–232.

Botts, H., & Patterson, J. (1987). Pension fund investments: An initial geographic assessment. *The Professional Geographer, 39*, 416–427.

Capital IQ Compustat. (2012). Compustat data. *Standard and Poor's*. New York, NY: McGraw-Hill Companies.

Carhart, M. (1997). On persistence in mutual fund performance. *The Journal of Finance, 52*, 57–82.

Choe, H., Kho, B., & Stulz, R. (2005). Do domestic investors have an edge? The trading experience of foreign investors in Korea. *Review of Financial Studies, 18*, 795–829.

Clark, G. (2005). Money flows like mercury: The geography of global finance. *Geografiska Annaler, Series B: Human Geography, 87*, 99–112.

Clark, G. (2010). Human nature, the environment, and behaviour: Explaining the scope and geographical scale of financial decision-making. *Geografiska Annaler: Series B, Human Geography, 92*, 159–173.

Clark, G. (2011). Myopia and the global financial crisis: Context-specific reasoning, market structure, and institutional governance. *Dialogues in Human Geography, 1*, 4–25.

Clark, R., Craig, L., & Wilson, J. (2005). *A history of public sector pensions in the United States*. Pension Research Council, The Wharton School of Business of the University of Pennsylvania. Philadelphia: University of Pennsylvania Press.
Coval, J., & Moskowitz, T. (1999). Home bias at home: Local equity preference in domestic portfolios. *The Journal of Finance*, *54*, 2045–2073.
Coval, J., & Moskowitz, T. (2001). The geography of investment: Informed trading and asset prices. *Journal of Political Economy*, *109*, 811–841.
Daniel, K., Grinblatt, M., Titman, S., & Wermers, R. (1997). Measuring mutual fund performance with characteristic-based benchmarks. *Journal of Finance*, *52*, 1035–1058.
Davis, P., & Steil, B. (2001). *Institutional investors*. Cambridge: The MIT Press.
DePew, C. (1895). *One hundred years of American commerce* (*Vol. 1*). New York, NY: D.O. Haynes and Co.
Dvořák, T. (2005). Do domestic investors have an information advantage? Evidence from Indonesia. *The Journal of Finance*, *60*, 817–839.
Everett, M. (1983). An extension of BLOC to valued graphs. *Social Networks*, *5*, 395–402.
Fleck, J. (1996). Informal information flow and the nature of expertise in financial services. *International Journal of Technology Management*, *11*, 104.
Forbes. (2012). Warren Buffett's stocks with the most insider buying. Retrieved from http://www.forbes.com/sites/gurufocus/2012/06/27/warren-buffetts-stocks-with-the-most-insider-buying/
French, K., & Poterba, J. (1991). Investor diversification and international equity markets. *American Economic Review*, *81*, 222–226.
Galbraith, J. (1961). *The great crash 1929*. Boston: Houghton Mifflin.
Geisst, C. (1997). *Wall Street: A history*. Oxford: Oxford University Press.
Global Finance. (2012). World's biggest banks 2010. Retrieved from http://www.gfmag.com/archives/130-october-2010/10634-worlds-biggest-banks-2010.html#axzz2A3Wixg94
Gong, H., & Keenan, K. (2012). The impact of 9/11 on the geography of financial services in New York: A few years later. *The Professional Geographer*, *64*, 370–388.
Goodwin, W. (1965). The management center in the United States. *The Geographical Review*, *15*, 1–16.
Gottman, J. (1970). Urban centrality and the interweaving of quaternary activities. *Ekistics*, *29*, 322–331.
Graves, W. (1998). The geography of mutual fund assets. *The Professional Geographer*, *50*, 243–255.
Green, M. (1983). The interurban corporate interlocking directorate network of Canada and the United States: A spatial perspective. *Urban Geography*, *4*, 338–354.
Green, M. (1993). A geography of institutional stock ownership in the United States. *Annals of the Association of American Geographers*, *83*, 66–89.
Green, M. (1995). A geography of US institutional investment, 1990. *Urban Geography*, *16*, 46–69.
Green, M., & Meyer, S. (1992). A temporal, spatial and functional view of investment companies in the United States. *The East Lakes Geographer*, *27*, 48–64. A copy of this article is available from the first author.
Green, M., Meyer, S., & Cromley, R. (1992). Canadian institutional and regional development. *Ontario Geography*, 38, 12–19. A copy of this article is available from the first author.
Green, M., & Semple, R. (1981). The corporate interlocking directorate as an urban spatial information network. *Urban Geography*, *2*, 148–160.
Griesst, C. (2012). *Wall Street: A history*. Updated edition. Oxford: Oxford University Press.
Grinblatt, M., & Titman, S. (1989). Mutual fund performance: An analysis of quarterly portfolio holdings. *Journal of Business*, *62*, 393–416.
Grinblatt, M., & Titman, S. (1993). Performance measurement without benchmarks: An examination of mutual fund return. *Journal of Business*, *66*, 47–68.
Gruber, M. (1996). Another puzzle: The growth in actively managed mutual funds. *The Journal of Finance*, *51*, 783–810.
Heath, C., & Tversky, A. (1991). Preference and belief: Ambiguity and competence in choice under uncertainty. *Journal of Risk and Uncertainty*, *4*, 5–28.
Hedgeco.net. (2013). Hedgeduation. Retrieved from http://www.hedgeco.net/hedgeducation/industry-overview.php
Hoover, E. (1948). *The location of economic activity*. New York, NY: McGraw-Hill.
Huberman, G. (2001). Familiarity breeds investment. *Review of Financial Studies*, *14*, 659–680.
Investment Company Institute (ICI). (2013). Investment company fact book. 3 2012 Facts at a glance. Retrieved from http://www.ici.org/pdf/2013_factbook.pdf

Investments and Income.com. (2013). History of pension plans. Retrieved from http://www.investmentsandincome.com/retirement-investments/history_of_pension_plans.html

Investopedia.com. (2013). Hedge fund. Retrieved from http://www.investopedia.com/terms/h/hedgefund.asp

Ivković, Z., & Weisbenner, S. (2005). Local does as local is: Information content of the geography of individual investors' common stock investments. *The Journal of Finance*, *60*, 267–306.

Jensen, M. (1968). Problems in selection of security portfolios: Performance of mutual funds in period 1945–1964. *The Journal of Finance*, *23*, 389–416.

Knoke, D., & Burke, P. (1980). *Log-linear models*. Thousand Oaks, CA: Sage.

Knox, P. (1995). World cities in a world system. In P. Knox & P. Taylor (Eds.), *World cities in a world system* (pp. 3–20). London: Routledge.

Latimer, M. (1932). *Industrial pension systems in the United States and Canada*. New York, NY: Industrial relations counselors.

Lewis, M. (2010). *The big short: Inside the doomsday machine*. New York, NY: W. W. Norton and Company.

Library of Congress. (n.d.). S.249 – Securities Act Amendment. Retrieved from https://www.congress.gov/bill/94th-congress/senate-bill/249

Malkiel, B. (1995). Returns from investing in equity mutual funds 1971 to 1991. *The Journal of Finance*, *50*, 549–572.

Malloy, C. (2005). The geography of equity analysis. *The Journal of Finance*, *60*, 719–755.

Markuen, A (1996). Sticky places in slippery space: A typology of industrial districts. *Economic Geography*, *72*, 293–313.

Marshall, A. (1920). *Principles of economics*. London: Macmillan and Co.

Martin, R., & Minns, R. (1995). Undermining the financial basis of regions: The spatial structure and implications of the UK pension fund system. *Regional Studies*, *29*, 125–144.

Minyanville.com. (2013). America's 10 oldest public companies. Retrieved from http://www.minyanville.com/special-features/articles/oldest-public-companies-america2527s-oldest-companies/9/23/2011/id/37022

Morgan, K. (2004). The exaggerated death of geography: Learning, proximity and territorial innovation systems. *Journal of Economic Geography*, *4*, 3–21.

Moriset, B., & Malecki, E. (2009). Organization versus space: The paradoxical geographies of the digital economy. *Geography Compass*, *3*, 256–274.

New York Times. (1905, November 19). The oldest life insurance company in the United States. *New York Times*.

Nussle, J. (2008, November 20). *Update of statistical area definitions and guidance on their uses*. Washington, DC: Office of Management and Budget.

O'Brien, R. (1992). *Global financial integration: The end of geography*. New York, NY: Council of Foreign Relations.

O'Hagan, S., & Green, M. (2004). Corporate knowledge transfer via interlocking directorates: A network analysis approach. *Geoforum*, *35*, 127–139.

O'Hagan, S., & Rice, M. (2012). The geography of corporate directors: Personal backgrounds, firm and regional success. *The Professional Geographer*, *64*, 586–601.

O'Hagan, S., & Rice, M. (2013). Exploring the personal histories of the top executives of US firms using a quantitative approach: Is there a geographical relationship with corporate headquarters, and does it influence firm performance?. *Geographical Research*, *51*. doi:10.1111/1745-5871.12004.

O'Hagan, S., Rice, M., & Green, M. (2008). North American corporate directors and educational affiliations: A geographical analysis. *Geography Research Forum*, *28*, 59–81.

O'hUallachain. (1985). Spatial patterns of foreign direct investment in the United States. *The Professional Geographer*, *37*, 154–162.

O'Hagan, S., & Green, M. (2002). Interlocking directorates: An example of tacit knowledge transfer. *Urban Geography*, *23*, 154–179.

Pearce, C., Green, M., & Baldwin, M. (2007). Developing habitat models for waterbirds in urban wetlands: A log-linear approach. *Urban Ecosystems*, *10*, 239–254.

Piggott, C. (2012, October 22). Managers stay close to their investments. *Financial Times Special Report, FTfm TheBanker*, p. 5.

Portes, R, & Rey, H. (2005). The determinants of cross-border equity flows. *Journal of International Economics*, *65*, 269–296.

QVM Group Perspectives. (2012). World capital markets – Size of global stock and bond markets. Retrieved from http://qvmgroup.com/invest/2012/04/02/world-capital-markets-size-of-global-stock-and-bond-markets/

Ramsay, I., & Blair, M. (1993). Ownership concentration, institutional investment and corporate governance: An empirical investigation of 100 Australian companies. *Melbourne University Law Review, 19*, 153–194.

Rice, M., & Semple, R. (1993). Spatial interlocking directorates in the Canadian urban system, 1971–1989. *Urban Geography, 14*, 375–396.

Sassen, S. (2001). *The global cities: New York, London, Tokyo* (2nd ed.). Princeton, NJ: Princeton University Press.

Schrader, S. (1991). Informal technology transfer between firms: Cooperation through information trading. *Research Policy, 20*, 153–170.

Securities and Exchange Commission. (2012a). *Official list of Section 13(f) securities*. www.sec.gov/divisions/investment/13flists.htm

Securities and Exchange Commission. (2012b). *Securities Exchange Act of 1934*. www.sec.gov/about/laws/sea34.pdf

Semple, R., & Phipps, A. (1982). The spatial evolution of corporate headquarters within an urban system. *Urban Geography, 3*, 258–279.

Shaw, M. (1981). Congress in the 1970s: A decade of reform. *Parliamentary Affairs, 3*, 272–290.

Smith, Y. (2010). *ECONned: How unenlightened self interest undermined democracy and corrupted capitalism*. New York, NY: Palgrave Macmillan.

Sobel, R. (1968). *The great bull market*. New York, NY: W.W. Norton.

Statistics Canada. (2011). Population and dwelling counts, for census metropolitan areas and census agglomerations, 2011 and 2006 censuses. Retrieved from http://www12.statcan.gc.ca/census-recensement/2011/dp-pd/hlt-fst/pd-pl/Table-Tableau.cfm?LANG=Eng&T=201&S=3&O=D&RPP=150

Stotz, O. (2011). The influence of geography on the success of private equity: Investments in listed equity. *Applied Financial Economics, 21*, 1605–1615.

SWF Institute. (2013). About sovereign wealth funds. http://www.swfinstitute.org/sovereign-wealth-fund/

Teo, M. (2009). The geography of hedge funds. *Review of Financial Studies, 22*, 3531–3561.

The Philadelphia Contributionship. (2013). Company history. Retrieved from http://www.contributionship.com/history/.

Thrift, N. (1994). On the social and cultural determinants of international financial centres. In S. Corbridge, R. L. Martin, & N. J. Thrift (Eds.), *Money, power and space* (pp. 327–355). Oxford: Blackwell.

Thrift, N., & Leyshon, A. (1994). A phantom state? The de-traditionalization of money, the international financial system and international financial centres. *Political Geography, 13*, 299–327.

Torrance, M. (2009a). The rise of global infrastructure through relational investing. *Economic Geography, 85*, 75–97.

Torrance, M. (2009b). Reconceptualizing urban governance through a new paradigm for urban infrastructure networks. *Journal of Economic Geography, 9*, 805–822.

Tversky, A., & Kahneman, D. (1974). Judgment under uncertainty: Heuristics and biases. *Science, 185*, 1124–1131.

US Census Bureau. (2012). *The 2012 Statistical Abstract*. Retrieved from http://www.census.gov/compendia/statab/cats/population/estimates_and_projections–states_metropolitan_areas_cities.html

Von Hippel, E. (1987). Cooperation between rivals: Informal know-how trading. *Research Policy, 16*, 291–302.

whalewisdom. (2011). *Whale wisdom: Track hedge funds using 13f filings*. Ashburn, VA.

Wheeler, J. (1990). Corporate role of New York City in the metropolitan hierarchy. *Geographical Review, 80*, 370–381.

Wheeler, J., & Mitchelson, R. (1989a). Atlanta's role as an information center: Intermetropolitan spatial links. *The Professional Geographer, 41*, 162–172.

Wheeler, J., & Mitchelson, R. (1989b). Information flows among major metropolitan areas in the United States. *Annals of the Association of American Geographers, 79*, 523–543.

Willekens, F. (1983). Specification and calibration of spatial interaction models: A contingency table perspective and an application to intra-urban migration in Rotterdam. *Tijdschrift voor Economische en Sociale Geografie, 74*, 239–252.

The geography of Canadian interlocking directorates: how do they relate to brain circulation?

Sean B. O'Hagan[a] and Murray D. Rice[b]

[a]*Department of Geography, Nipissing University, North Bay, Canada;* [b]*Department of Geography, University of North Texas, Denton, USA*

This study examines the spatiality of interlocking directorates of Canadian companies, with particular emphasis given to international interlocks. From a geographical standpoint, do these connections to foreign firms reinforce the importance of Canada's top corporate cities as 'gateways' to international business? Is there a geography associated with the interlocking to these gateway cities? Findings build on James O. Wheeler's body of research focused on corporate headquarters. This study is placed within the context of the current economic geography dialogue. Particular focus is placed on Saxenian's theory of brain circulation, wherein she argues that individuals with international experience serve as brokers linking domestic businesses with those in distant regions. A further purpose of this study then is to examine Saxenian's argument in the context of Canadian corporations. First, is there a geographical association between directors' foreign educations and where Canadian firms are interlocked to internationally? Second, the corporate performance of Canadian firms interlocked internationally will be measured against those firms that are not. It seems logical to assume that increased connections would yield more diverse and rich knowledge, which could lead to better decisions, and thus corporate performance, in today's global world. But does it? This study finds that being linked to an international company through a common director did translate into more profitability for Canadian firms. Findings also revealed a geographical link between international interlocks and a director's foreign education. However, when Canadian firms were interlocked with an international company with a director who had international education, greater profits did not necessarily result.

Introduction

In an increasingly competitive business environment, profitable firms find ways to secure competitive advantages. An example of this is knowledge in the form of corporate strategy. How do high-performing firms constantly upgrade their knowledge to improve corporate strategy and sustain growth? One generally accepted method is through interlocking directorates. An interlocking directorate occurs when a director sits on two or more boards, creating a link between the two boards (and firms). A characteristic of an interlocking directorate is the fluidity of knowledge transfer the position facilitates as a director is able to pass knowledge of one from to another. Resource dependency theorists argue that these connections yield knowledge transfer between companies (Pfeffer & Salancik, 1978). One purpose of this study is to examine the spatiality found within the interlocking of Canadian companies. Particular emphasis is given to international interlocks, that is, when Canadian companies are linked to international companies through the same director. From a geographical standpoint, cities are important nodes in a globalizing world. McCann and Ward (2011) suggest that an important area of study is to focus on the practices that constitute cities as sites of intersection between network typologies and territorial legacies. In response we ask, do interlocks to foreign firms reinforce the importance of Canada's top corporate cities as "gateways" to international business? Is there a geography associated with these international connections and gateway cities?

This study extends the work of James O. Wheeler's headquarters location research (Holloway & Wheeler, 1991; Wheeler, 1985, 1986, 1988a, 1988b, 1990), as well as his information flows research (Mitchelson & Wheeler, 1994; Wheeler, 1999; Wheeler & Mitchelson, 1989). Findings are placed within the context of the current economic geography dialogue that examines the relationship between geography and knowledge in today's global economy. Particular focus is placed on Saxenian's (2005, 2006) theory of brain circulation, wherein she refers to the circular movement of skilled labor across geographic space. Saxenian (2002) argues that individuals with international experience serve as brokers linking domestic businesses with those in distant regions and that the circular movement of these individuals brings intangible knowledge that proves invaluable for the competitiveness of firms and the regions where they are located. A further purpose of this study is to examine her brain circulation theory in the context of Canadian corporations. This is accomplished by examining two questions. The first looks at whether there is an association between directors with a foreign education and where Canadian firms are interlocked internationally? In other words, is a director's education and where she/he is interlocked internationally correlated? The second question analyzes the association between Canadian firms interlocked internationally with corporate performance. These connections are an excellent example of network typologies influencing territorial legacies. While a range of factors impact firm performance, director knowledge of the external environment should be a significant one. In our view, exposure to international institutions and linking to international businesses yields more diverse and rich knowledge that could lead to better decisions, and thus corporate performance, in today's global world. Yet the relationship is a nuanced one. This study investigates whether being linked to an international company through a common director translates into higher levels of profitability for Canadian firms. An important part of this geographical analysis is the study of geographical connection between international interlocks and a director's foreign education. However, before the article addresses the details of this analysis, it opens with a discussion of the salient literature.

Knowledge its transfer and geography

For many companies today, knowledge is considered the most strategically important resource. Some individuals like Drucker (2005) go as far to suggest it is the only meaningful resource with continuous generation necessary for companies to remain competitive. He argues this because knowledge allows firms to capitalize on their traditional resources to create new processes and innovative products. Chirico and Salvato (2008, p. 169) are representative of the knowledge supporters position when they argue that a knowledge-based economy has materialized and acknowledge that the enablers of firm acclimatization are crucial to understanding competitive advantage.

For academics, differentiating between internally produced knowledge and that acquired externally is an important distinction. Internally procured knowledge through a company's research and development department, always historically important, is advantageous because of its exclusivity to that company. Being firm-specific makes it problematic for competitors to duplicate, especially if it is tacit knowledge. Unfortunately, it is also expensive to generate. Consequently, to continually innovate in today's rapidly paced environment, externally acquired knowledge has become increasingly important. Accessing external knowledge allows companies to adapt quickly to changes in products and processes, an important concept in today's global economy.

Nonaka and Takeuchi (1995) highlight knowledge management as a strategic asset used by an organization to improve performance. They argue that corporations need to identify conduits of knowledge facilitated through individuals' insights and experiences. Through collective learning, this knowledge ultimately crosses departmental and divisional lines, thereby creating a competitive advantage. When individuals are crucial to firm success, such as boards of directors, then knowledge acquired by these actors outside the firm should make the organizations they administer more competitive.

With the increasing importance of external knowledge to firm success, academics have studied a number of attributes regarding this phenomenon. Economic geographers have focused on the spatiality of its transfer, with two opposing schools of thought emerging. The first approach focuses on the sharing of knowledge between geographically proximate economic actors, termed learning regions, innovation systems, clusters, and so on (see Bathelt, Malmberg, & Maskell, 2004; Cumbers & MacKinnon, 2004; Hassink, 2005). This research originally dates back to Marshall (1920) who suggests that firms in spatial agglomerations benefit from technological spillovers. Ever since, academics have acknowledged regional context as an important source of intangible assets necessary for competitive advantage.

A key basis for this line of reasoning is that learning occurs more easily at the regional level because knowledge is grounded in spatial connections and institutions (Amin, 2004; Storper, 1997). Institutions are mechanisms of social order governing the behavior of individuals. The term normally includes formal structures at a spatial scale such as legal rules, property laws, government policies, as well as informal habits, codes of conduct, and organizational cultures. Within this line of reasoning, Granovetter's (2005) embeddedness-network paradigm argues that relationships between individual actors are set within social networks. Since regional actors share these institutions, geographers argue that these place-based dynamics provide a superior avenue for knowledge to be passed.

When two individuals share a similar background, the likelihood of knowledge transfer is enhanced. Recent research, as seen in Malmberg and Maskell (2010), Maskell and Malmberg (2007), Maskell (2009), and Morgan (2007), claims that physically proximate firms interact to create innovative processes and products. Areas of specialization with their high levels of direct face-to-face relations support a commitment towards collective learning and shared prosperity. While explicit knowledge moves easily and quickly, impediments to tacit knowledge transfer are for the most part still intact. Face-to-face communication is still necessary for its transfer. Thus, the characteristics of tacit knowledge, that is the knowledge that leads to competitive advantage, is best transferred at the regional level. As individuals relate to one another, trust develops. This in turn encourages the diffusion of technical knowledge and skills within the area. Since the environment is conducive to the diffusion of knowledge, the concept of place plays a pivotal role in its transfer.

The second approach to knowledge transfer does not deny the importance of regional interaction but argues that neglecting areas beyond the local level as sources of knowledge is misguided (Amin & Cohendet, 1999; Jones, 2008). Even in industries that exhibit an extreme regional focus, an increasing number of firms are linked to firms outside of the locale (Ter Wal & Boschma, 2009). For example, not all relevant firms in the entertainment business are located in Hollywood. Indeed an increasing number are located outside of it even when the region holds preeminence in the industry. Undeniably, the production and dissemination of knowledge is increasingly occurring on a global level. Therefore, knowledge is not fixed in a specific territorialized set of connections (Vallance, 2007); thus, it is important to examine these typologies (McCann & Ward, 2011).

As recognized by Malecki (2010a), knowledge exists within a network of decentralized and isolated areas. To utilize these different sources of knowledge, a company must use its connections to become a local throughout the globe, positioning itself to utilize the appropriate knowledge from these various sources. This argument suggests that interregional linkages are necessary because they provide new knowledge unavailable in the local region. If networks occur between "actors" of different regions, then unique experiences can act as a balance to local ways of conducting business (Malecki, 2010b).

Exposure to external knowledge avoids an over-reliance on regional knowledge, which could prevent firms from adapting to long-term changes (Green & Semple, 1981; O'Hagan and Rice, 2012; Rice, Tierney, O'Hagan, Lyons, & Green, 2012). It is important to acknowledge that firms in any given region may initially be highly competitive. However, over time, these firms may become less innovative because they limit their competition energies and focus to their region and not the entire global industry. Folta, Cooper, and Baik (2006) argue that clusters result in better firm performance, but they also suggest that diseconomies of agglomeration play an increasingly important role as these regional groupings of firms grow. To develop this thesis, Amin and Cohendet (1999) discuss an ideal-type structure of a global firm, one in which the core knowledge-intensive activities are divided between divisions located in dissimilar environments and that draw on their own unique knowledge.

As some economic geographers recognize knowledge transfer beyond a regional approach, the need arises to explore alternative ways of conceptualizing knowledge movement, especially outside the locale. These involve studying knowledge flows between firms through different types of territorially unbounded network formation. This need is related to Castells' (1996, p. 446) influential work regarding networks of knowledge in today's post-modern society. He hypothesizes that space of flows are the result of specific micro-networks within the larger system.

In their examination of policymaking of urban environments in a globalized world, McCann and Ward (2011, p. xvii) provide a similar argument to Castells. They call for more empirical accounts to illustrate the "struggles, practices, and representations that underpin urban–global relations and that assemble or territorialize global flows." Robinson's (2011) response to their argument is noted later in the same text. She suggests that the potential for policy ideas to circulate in different directions across the globe lies in the notion that agents of policy, whose ambition is substantially transnational, exist. Her point lies with individuals directly influencing urban policy, and thus urban competitiveness, in a global world. However, she also identifies that the concept of global does not signify a single planetary circuit of knowledge but, rather, overlapping and multiple transnational tracks of engagement, influence, and information. It is our argument that similarly directors of corporations play a significant role in urban competitiveness. In a global world, this influential cohort is a pipeline of international knowledge transfer amongst corporations, especially if they are a part of interlocks that transcend national boundaries. These individuals ultimately influence urban competitiveness by directly influencing the companies that lie within urban environments.

Alliances, foreign direct investment, acquisitions, and mergers are authorized agreements that endorse inter-firm knowledge transfer locally and globally. Less sanctioned, but still very important knowledge networks include links to suppliers and buyers, start-ups by individuals belonging to another firm, staff movement, and rumors. The focus of this study is on another informal knowledge network interlocking directorates.

Directors involved in an interlock are important because they hold an extensive amount of knowledge about the corporate landscape. At the individual level, directors involved in an interlock are part of a larger social network associated with the companies they sit on. Castells (1996) specifically highlights the importance of the managerial elite to the spatial study of knowledge flows. Castells (1996) and Sklair (2001) argue that this group of people has a disproportionate influence on the macro-network of knowledge transfer. The personal spatial flow of knowledge amongst the micro-networks of the managerial elite, in this case through interlocking directorates, is worthy of study. In response, this study takes up McCann and Ward's challenge within the context of today's economic geography literature. Additionally, this study adapts Wheeler's headquarters research as well as his information flows research to examine today's urban–global relationship.

The knowledge economy, interlocking directorates, and brain circulation

It has been established that a critical challenge for management to ensure sustained growth is to procure external knowledge. One of the most important ways for a firm to access external knowledge is through its board of directors (Connelly, Johnson, Tihanyi, & Ellstrand, 2010). Considered informed and accomplished professionals, directors provide independent knowledge on the external economic climate. Being removed from every day activities of a firm, directors offer fresh ideas for a firm's strategy. Perhaps the most important way a board member can obtain knowledge is to serve as a board member of another firm. Termed an interlocking directorate, sitting on two or more boards allows an individual to pass knowledge of one firm to another.

Research on interlocking directorates has developed extensively over the last three decades. Dominated by sociologists, these studies have attempted to explain why these connections take place. As the business environment has changed, different paradigms have arisen to explain this phenomenon. From this literature, resource dependency has emerged as the dominant theory to explain why interlocking directorates occur (Pfeffer & Salancik, 1978). This theory states that firms are unable to generate all the necessary inputs for production. Firms respond with relationships to external elements of the environment to ensure a supply of required resources. By connecting to external sources, knowledge of the environment is gained and uncertainty decreased.

Obtaining knowledge is difficult because of its embedded nature within the firm's mechanisms and process, as well as within the firm's personnel. However, a director sitting on the board of two firms allows knowledge about corporate policy to be transferred (Malmberg, Solvell, & Zander, 1996; Pennings, 1980; Singh, 2007; Strauss, Nevries, & Weber, 2010). Returning to the different types of knowledge, an important advantage of interlocking directorates is that this knowledge transfer does not only occur between geographically proximate firms. In fact, tacit knowledge, the knowledge that leads to competitive advantage, can be transferred via an interlock between distant firms.

Geography of interlocking directorates

As geographers, we suggest that the investigation of asking why interlocks occur addresses only a portion of the overall phenomenon of corporate interlocking. By adding a spatial component, we are able to ask: Where do interlocking directorates occur and does this influence the profitability of firms? From this generative, locus one can then suggest reasons why interlocking directorates provide a basis for knowledge transfer. It can only be assumed that the increased knowledge introduced by a director enables more knowledge that decreases firm uncertainty and leads to better business decisions.

Green (1980) and Green and Semple (1981) were the first researchers to examine the geography of interlocking directorates and its consequences. They hypothesized that the United States manufacturing belt was dominated by intraregional interlocking. This was advantageous when firms of the manufacturing belt were on the cutting edge of technological development. However, over time, not interlocking with firms in other regions deprived north eastern firms of vital information and led to the region's decline.

Rice and Semple (1993) examined interlocks spatially in support of a different purpose. Using interlocks, they found that interlocks concentrated corporate control in Toronto more so than previously thought. Temporally, Montreal decreased in corporate potency to become a regional center in eastern Canada, and Calgary emerged to become a regional center in western Canada.

Kono, Palmer, Friedland, and Zafonte (1998) found that when local and nonlocal interlocks were aggregated into a same sample, the presence of exclusive upper-class social clubs in a corporation's headquarters city did not influence corporate interlocking. However, an interesting result occurred when the sample was dissected into local and nonlocal links. It was found that when corporations were headquartered in cities with exclusive upper-class social clubs, they were more likely to maintain local interlocks. On the other hand, corporations not headquartered in cities with exclusive upper-class social clubs were less likely to maintain local interlocks.

O'Hagan and Green (2002a, 2002b, 2004) explored the gap between previous resource dependency-knowledge transfer research and geographical findings. Two notable findings are worth mentioning here. First, a spatial component was added to the resource dependency paradigm. Second, using a Poisson regression model and network analysis, components of a city that initiate and attract interlocking, and thus knowledge transfer, were identified. O'Hagan and Green found that geography was more important in the American network than in the Canadian network.

As economies of the world have become more interlinked, research has attempted to capture these changes. van Veen and Kratzer (2011) examined interlocking directorates across European countries and found that different countries took quite dissimilar positions. They found a link between the structure of the interlocking directorate network within a country and how that country was positioned in the international network. With a more intercontinental focus, Kentor and Jang (2004) studied the Fortune Global 500 between 1983 and 1998 and found a significant increase in the total number of interlocking directorates and even greater growth in transnational interlocks. Additionally, they found that the growth was primarily between firms within Europe as well as European countries and the United States.

Interlocking directorates provide a unique setting for studying the sharing of knowledge. Over time, relationships are built up not only with other directors in the same urban area, but also with directors in distant locations. While local relationships are important, it must be stressed that nonlocal connections also play a significant role in sustaining competitiveness. These connections provide additional sources of knowledge about producers, consumers, and regulators. This study will provide a bridge to address the gap between the resource dependency-knowledge transfer research and geography. Most importantly, the study proposes to bring Saxenian's theory (2002), which utilizes a metaphor of brain circulation, into the conversation on the subject.

Brain circulation

Saxenian (2002, 2006) and Kuznetsov (2006) propose the concept of brain circulation to explain when people emigrate from their home country and return later once they have increased their knowledge and skill base. Brain circulation is sometimes referred to as the circular movement of skilled labor across nations. It builds on the concept of brain drain, which gained popularity as skilled labor emigrated from developing nations to other countries in search of better opportunities. In the 1990s and onwards, many of these individuals returned to their home countries. The current economic strength of countries like India, China, and Taiwan are partly accrued by this circular movement of skilled labor. Researchers such as Saxenian and Kuznetsoz argue that this circular movement brings intangible knowledge back to the home country that proves invaluable for growth.

The concept of brain circulation is applicable to knowledge connections and generation at a regional level, but it extends this region-oriented research to the international environment. As early as 40 years ago, Zelinsky (1971, p. 230) proposed "the hypothesis of the mobility transition." In it, he prophesied the "significant international migration or circulation of [a] skilled and professional person." More typical of transnational communities in today's global economy, Saxenian argues that individuals with international experience serve as brokers linking domestic businesses with those in distant regions. Further, she argues that the circular movement of these individuals brings intangible knowledge that proves invaluable for the competitiveness of firms and the regions where they are located. She states (2002, p. 30) "a scarce resource in this new environment is the ability to locate foreign partners and manage complex businesses across cultural and linguistic lines." Jöns (2009) extends the concept to include individuals associated with higher education and research. Examining the research residencies of visiting scientists and scholars in Germany, he argues these networks create the construction of global knowledge nodes. We argue that examining the circular movement of boards of directors, individuals who are key to corporate success, provides fertile ground to test Saxenian's hypothesis. The purpose of this study is to examine her theory in the context of Canadian corporations and the cities where they are located. First, from a geographical standpoint, this study questions whether Canadian companies possessing directors with international links reinforce the importance of Canada's top corporate cities? Second, we examine whether interlocking to international firms influences Canadian firms performance. Following the research of Kuznetsov and Sabel (2006), it is noted that these interlocks bridge institutions between countries as well as the risk-mitigating knowledge associated with institutions that these individuals possess is invaluable for firm prosperity and the cities where they are located.

Data

To examine interlocks, a thorough director data set as identified by the Financial Post *Directory of Directors* for the year 2011 was used. The register of companies includes the names of both publicly traded and privately owned firms, along with their addresses in addition to the names of their executive officers and directors. Criteria for inclusion of companies in this source include incorporation in Canada; substantial revenue or assets; and Canadian residency for the majority of the directors. Once a company qualifies for inclusion, its officers and directors automatically meet the criteria for a personal listing. The directory provides information on the educational attainments of only a portion of directors. The educational data is self-reported and some directors choose not to reveal their degrees. Some directors, of course, may not have any university education, but this would be an unusual occurrence.

The data set includes 18,298 businesswomen and men showing the directorships and offices they hold for 5,548 companies. Of the directors, 4,702 were interlocked, sitting on the boards of at least two firms. Since many were a part of more than one interlock, 15,654 examples of interlocks occurred. Of the companies examined, 5,105 were Canadian and 443 were international. Foreign companies only remained part of the data set if they possessed a director that was interlocked to a Canadian company. This resulted in 682 directors being interlocked to a foreign company. Again, since many were part of more than one interlock, 1,257 total interlocks to international firms occurred. Education was selected as a proxy for personal experience as this data is readily available. The location of a director's education plays a pivotal role in where she/he accumulates business knowledge as well as where she/he has personal network connections. Consequently, it notes where directors have previous experience. A limitation of this study is the exclusion of previous work experience as this would also indicate where directors obtained previous knowledge and where they built up their social networks. Unfortunately, this data was not readily available. Financial data was obtained from COMPUSTAT, a database of financial, statistical, and market information on active and inactive global companies throughout the world. From this, gross profit (loss) was selected as a proxy for firm performance. Here, gross profit is the difference between revenue and cost, such as making the product or providing a service, deducting overhead, payroll, taxes, etc. Importantly, earnings per share, is another variable whose results are similar to those reported for gross profit but are not incorporated into this paper. Of the interlocks, educational data for the director is known in 10,779 cases while financial data is known in 2,977.

Results

Our results are presented in two parts. First, using descriptive statistics, a general overview of the relationship between firm headquarters and interlocking is provided. In essence, the initial results reported here are from a geographical perspective. Table 1 provides city-level data, summarizing the number of headquarters they house, the number of directors that sit on these firms' boards, and the interlocks for which they are a part of.

In the first portion of Table 1, headquarter results generally correspond to research findings on the Canadian corporate hierarchy (Rice, 2006). In that hierarchy, Toronto is the most important center with Vancouver, Calgary, and Montreal following as an important second tier. Perhaps, Montreal housing the fourth largest number of head-quarters was the most surprising as this city continues to diminish in prominence in Canada's corporate hierarchy. While the top tier of cities maintains a disproportionate number of headquarters compared to the rest of Canada, the number of directors located in these largest cities is even more pronounced when compared to small- and medium-sized cities. Additionally, these directors are more likely to be part of an interlock and important to this study as part of an international interlock. These four cities possess 77% of total headquarters; 77% of directors; 80% of interlocks; and 87% of international interlocks. The list of top 20 cities presented here reveals corporate concentration further, by

Table 1. List of headquarter cities, number of directors, and number of interlocks in each city.

	Headquarters #		Directors #		Interlocks #		Inter. Interlocks #	
City	#	%	#	%	#	%	#	%
Toronto	1701	29.8	7689	31.8	4162	31.9	404	33.6
Vancouver	1231	21.6	3599	16.2	2413	19.5	282	23.5
Calgary	875	15.3	3932	16.2	2212	16.9	182	17.1
Montreal	563	9.8	3106	12.8	1512	11.5	116	12.6
Edmonton	146	2.5	673	2.7	312	2.3	28	2.3
Winnipeg	129	2.2	646	2.1	335	2.0	26	1.1
Ottawa	102	1.7	576	1.3	232	1.2	31	1.5
Quebec City	82	1.4	345	1.4	160	1.2	6	0.5
Halifax	81	1.4	345	1.4	183	1.4	8	0.6
Saskatoon	49	0.8	204	0.8	101	0.7	10	0.8
Kitch-Wat-Cam.	45	0.7	311	1.2	117	1.9	18	1.5
Regina	42	0.7	221	0.9	128	0.9	1	0.0
London	39	0.6	151	0.6	67	0.5	7	0.2
Victoria	36	0.6	140	0.5	63	0.4	1	0.0
Burlington	26	0.4	145	0.6	55	0.4	4	0.3
Kelowna	26	0.4	108	0.4	47	0.3	0	0.0
Hamilton	25	0.4	81	0.3	38	0.2	4	0.3
Guelph	24	0.4	83	0.3	42	0.3	1	0.0
Moncton	24	0.4	150	0.6	73	0.5	2	0.1
Red Deer	24	0.4	120	0.5	44	0.3	1	0.0
St. John's	22	0.3	83	0.3	43	0.3	2	0.1

accounting for 93% of the headquarters; 94% of the directors; 96% of the interlocks; and 98% of the international interlocks.

These four cities generally follow a pattern of increasing concentration with each variable. Moving beyond previous findings that use the brick and mortar headquarters locations as a proxy for corporate control, Table 1 suggests an even more concentrated control of decision-making in a few gateway Canadian cities. The results show how some companies, and the cities where they are located, are more interconnected through interlocked directors than previous research suggests. Thus, Robinson's (2011) argument for a need to develop a spatial vocabulary more adequate to capturing the spatialities of circulation is relevant here. Her focus is on topological space rather than physical distance. Specifically, she refers to topological accounts of space demonstrating how people, and their ideas, can be closer in relationships and proximity than physical distance leads us to believe. This reinforces this study's argument that more interconnected firms, and their cities, are more competitive in today's global economy.

International corporate knowledge flows through the top corporate hierarchy of Canadian cities more than previously reported. Our analysis of the interlocking network suggests that existing research on the Canadian corporate city structure has neglected to examine the ultimate decision makers for firms. This argument maintains that the competitiveness of cities is determined not only by the number of firms they contain, but are also shaped by the nature and number of social ties to other major firms in other cities.

Addressing the concerns of the second question, Figures 1–4 illustrate if geographical patterns are associated with the international interlocking of Toronto-, Vancouver-, Calgary-, and Montreal-based firms. Figure 1 shows Toronto as the city at the top of the Canadian corporate hierarchy. It reveals no particular geographical pattern with

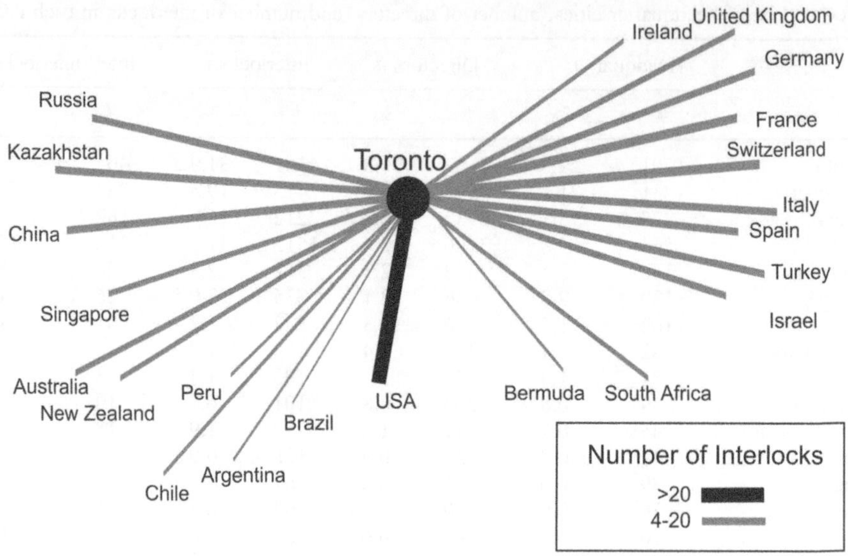

Figure 1. International interlocks with Toronto-based firms, country level.

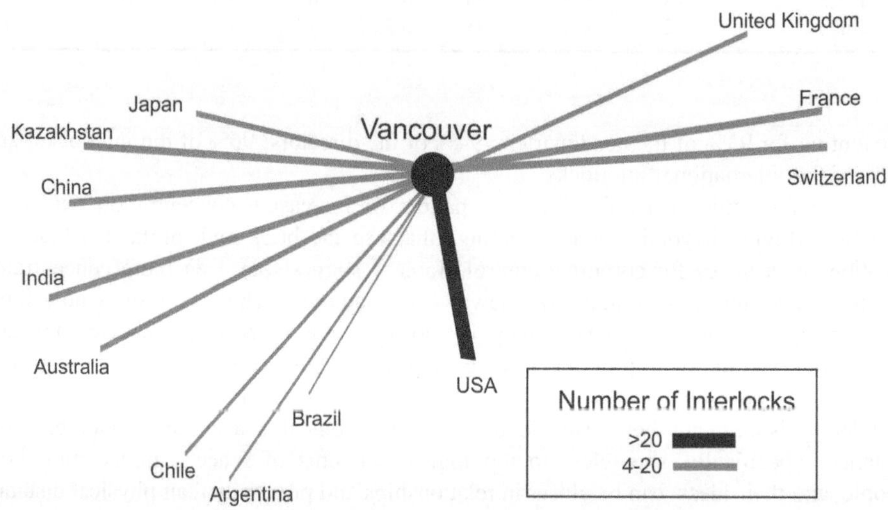

Figure 2. International interlocks with Vancouver-based firms, country level.

international interlocks dispersed throughout the world. One important distinction is the heavy reliance on interlocking with American corporations. While not a top tiered global city, the position of Canada's most internationally recognized corporate city conforms to previous research findings on the subject. Toronto firms possess close ties to their nearest neighbor, but linkages are generally dispersed throughout the world.

Vancouver, an emerging corporate city in western Canada, is also strongly linked to American firms. As the Canadian economy is closely tied to that of the United States' economy, this result is not unexpected. In fact, American linkages dominate for all four of

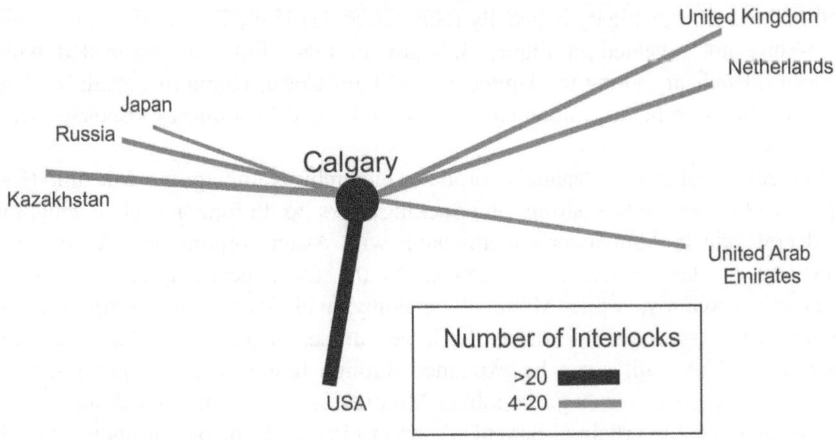

Figure 3. International interlocks with Calgary-based firms, country level.

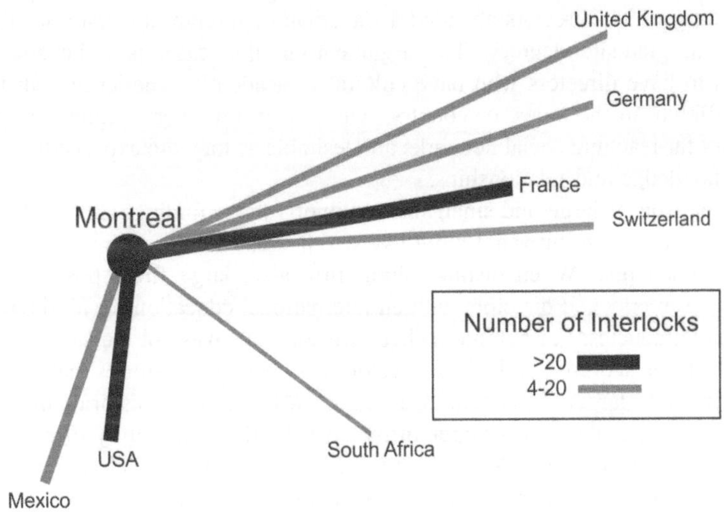

Figure 4. International interlocks with Montreal-based firms, country level

Canada's top cities. However, Vancouver shows a robust connection to Asia. These findings indicate that if a director sitting on a Vancouver firm's board also sits on an international company's board of directors, it is likely that the company is located in Asia. Fewer links were found to exist with European firms. While the concept of distance plays an important role in linking Vancouver to Asian firms, culture is also an important factor as well. A dramatic increase in immigration since the 1980s has resulted in over half the population of Vancouver not speaking English as their first language, according to Statistics Canada (2006). Today, almost 30% of the city's inhabitants originate from different Chinese ethnicities. The second largest cultural group in Vancouver is Indian (5.7%) and fourth largest is Japanese (2.0%), also reflecting an Asian presence in Figure 2 (Statistics Canada, 2006).

Calgary, the third ranked city in Canada's corporate hierarchy, reveals limited geographical patterning beyond the United States. In more of an infancy stage compared to the other three cities, Calgary is heavily reliant upon the United States for its international links. While not revealed in Figure 3, many of these links are associated with the American oil industry along the United States' Gulf Coast. Home to Canada's oil sector companies, linkages persist to companies that are located in countries associated with the oil industry.

Montreal's position in Canada's corporate hierarchy is undergoing a decline (Rice & Semple, 1993). The city's strong interlocking links to Europe reveal a geographical dependency akin to Vancouver's relationship with Asian corporations. A geographical dependency can be positive and negative. As the Asian economy rises, so too does Vancouver's economy, while Montreal economy, with its ties to Europe's uncertain economy, declines. One final note is the large number of links from Montreal firms to French firms. This outlier can be explained through linguistic and cultural similarities between the two countries as metropolitan Montreal is 65% French-speaking.

As a focus of this article is Saxenian's brain circulation theory of knowledge flows, Table 2 provides a broad geographical comparison of interlocked and noninterlocked directors' educations. The findings suggest that interlocked directors are more likely to have an international education than noninterlocked directors. Of additional interest is the data regarding which directors attended a Canadian or foreign university for their undergraduate and graduate degrees. The argument, in this case, is it benefits Canadian companies to have directors who have cultural or academic experience with institutions that are different in the cities, or country, where they now work. Additionally, directors that possess far-reaching social networks are desirable as they are exposed to a network of external knowledge and relationships.

A comparison of large and small firms supports these findings. Interlocked directors for both small and large firms are more likely to possess an international education rather than a Canadian one. When distinguishing firm size, large firms possessed a greater percentage of interlocked directors with an international education for all three categories of university graduates. For noninterlocked directors, this was not the case. This indicates that smaller firms acquire interlocked directors with personal histories that are geographically closer while larger firms acquire directors with personal histories that are further away. It can be speculated that larger firms have the financial ability to conduct a more extensive search or are better known across a broader geographical area so as to entice individuals from more distant locations. Interestingly, directors obtaining an undergraduate degree in Canada, and a graduate degree from an international institution were quite high, in proportion, especially for large firms. The majority of these international

Table 2. Percentage of directors with international education (statistics in parenthesis exclude education received in the United States).

Education level and size of firm	Interlocked director	Noninterlocked director
Undergraduate	22.3 (8.0)	14.9 (5.9)
Large firms	26.2 (8.2)	17.0 (7.0)
Small firms	19.1 (6.3)	12.8 5.5
Graduate	31.1 (10.0)	22.5 (9.7)
Large firms	32.1 (9.3)	21.0 (9.8)
Small firms	30.2 (7.4)	23.4 (6.9)
Undergrad Canada but Grad not Canada	19.9 (5.6)	10.4 (4.8)
Large firms	24.6 (6.1)	9.9 (5.1)
Small firms	13.2 (4.9)	10.6 (4.7)

educations were obtained from American universities, with United Kingdom-educated directors following a distant second. This can be explained by the large number of graduate opportunities in the United States compared to Canada, especially those associated with prestigious MBA programs. More importantly, large companies have the financial capability, as well as other inducements, to pursue graduates of these prestigious American graduate programs. The salient point here is that by acquiring directors with more international personal histories, large firms acquire directors who offer knowledge pipelines to distant locations. This places large firms in a superior situation to adapt their corporate strategies.

Due to the economic and cultural relationship between Canada and America, this study separated American data from the data containing other foreign interlocks and educations. Canada and the United States are currently the world's largest trading partners, with the friction of distance between the two countries much lower than that between Canada and other countries. Thus, as noted in Table 2, results that exclude educations obtained in the United States are provided in parenthesis. Results support this study's argument that a large portion of the international degrees obtained by directors were obtained in American institutions. While patterns are similar to those already reported, a dramatic decrease occurs in directors obtaining an international education.

These results are of interest when a comparison is made between where Canadian companies are linked to and where directors of those particular interlocks attended university. If interlocked internationally, a remarkable 79.6% of undergraduate alumni and 80.1% of graduate alumni are part of an interlock to a firm in the same country as they attended university.

When breaking firm size into small and large, a larger percentage of directors from large firms are part of an interlock to a firm in the same country as they attended university. A remarkable 86% of undergraduates of large firms attended a university in the same country where they sit on a board of directors. In other words, they link the Canadian company whose board they sit on to a firm in the same country they attended university. A by-product of possessing a director with this type of background is that the Canadian company is exposed to individuals that have direct corporate knowledge in the same foreign country, whether intended or not. This noteworthy correlation substantiates Saxenian's concept of brain circulation. Individuals exposed to institutions and connections in a foreign country through education or past experience are likely to sit on a board of directors of a company from that country in addition to a Canadian one. Individuals with international experience appear to act as brokers linking Canadian businesses with businesses in foreign countries. In Table 3, American interlocks were excluded for results

Table 3. If interlocked internationally, percentage of directors linked to a firm in the same country as they attended university (statistics in parenthesis exclude interlocks with American United States).

Education level and size of firm	Interlocked director
Undergraduate	79.6 (28.4)
Large firms	86.3 (30.0)
Small firms	73.9 (26.6)
Graduate	80.1 (27.9)
Large firms	82.9 (31.25)
Small firms	78.4 (23.0)

in parenthesis. Again, a dramatic decrease occurs, indicating the interconnectedness of the American and Canadian interlocking network. This closer examination of the data indicates that when directors sit on a Canadian firm as well as an international firm from other than the United States, many of these directors attended an American university.

Using inferential statistics, we examine if firms possessing directors that are linked externally, and in particular internationally, translates into greater profitability. In essence, we want to build upon results thus far and further test Saxenian's theory. Research findings on this subject are mixed. Fligstein and Brantley (1992) found a negative association between interlocks and profitability among large American firms. On the other hand, Pennings (1980), Carrington (1981), Burt (1983), and Baysinger and Butler (1985) found generally positive associations between interlocking and profitability. Using the director as the unit of analysis, Table 4 provides output from four *t*-tests. *t*-test 1 suggests that directors who are a part of an interlock are more likely to sit on companies with higher profits (or less loss). This result makes sense as the increased connections yield better decisions.

t-test 2 and *t*-test 3 examine directors interlocked internationally or those that attended an international university. *t*-test 2 compares directors of Canadian firms interlocked to foreign companies with Canadian companies interlocked to domestic ones. Results indicate that directors who were a part of an international interlock were more likely to sit on Canadian firms' board that experienced higher profits (or less loss). *t*-test 3 compares interlocked directors who attended an international university to those interlocked directors who attended a Canadian university. Here, profitability results of Canadian firms are not as strong. While the mean for profitability suggests that interlocked directors that attended an international university were higher, and thus agree with our hypothesis, they were not statistically significant.

The fourth *t*-test is a more rigorous to an assessment of Saxenian's brain circulation theory. Broken into two samples, the first group encompasses directors that attended a foreign university and now sit on the board of directors for a company in that same country (as well as sitting on a board of a Canadian company). The second group is composed of all other directors. Importantly, the fourth *t*-test is solely comparing performance of Canadian firms in the two samples. Similar to Model 3, the mean for profitability suggests that directors that attended both an international university and Canadian

Table 4. *t*-tests for gross profit.

Model	Sample	*F*	Sig.	Mean
Model 1	Directors of interlocked companies	0.017	0.034	305.44
	Directors of non-interlocked companies			228.38
Model 2	Directors of a Canadian and international company	5.352	0.021	387.94
	Directors only of Canadian companies			269.08
Model 3	Directors with international education	0.356	0.551	352.86
	Directors with Canadian education			207.54
Model 5	Directors with both international and Canadian education	0.11	0.741	705.37
	Directors with Canadian education			369.41
Model 4	Directors with international education (excluding US)	4.037	0.045	1458.93
	Directors with Canadian or US education			1107.57
Model 6	Directors with international education and ether Canadian or US education	5.114	0.024	2018.37
	Directors with Canadian or US education			1107.57

university were higher. Therefore, these results agree with our hypothesis. However, they were not statistically significant.

The fifth and sixth *t*-tests again modified the data in order to filter out the interconnectedness of the Canadian and American economies. As with the previous *t*-tests, directors with an international education (in this case excluding alumni of American universities) translated into greater profits for their corporations. These results were statistically significant. The results suggest that firm performance is enhanced when corporations hire directors with educations from countries other than the United States and Canada. Greater profits occur in firms that maintain directors exposed to further international experiences and who create global social networks.

The next step of this study was to utilize the firm as the unit of analysis. The multivariate results in Table 5 show that a relatively small percentage of the variance in firm profitability is accounted for by the number of directors and the spatiality of their interlocks (ranging from .37 to .41). This was expected as this is but one factor of firm knowledge gathering. As suggested earlier, alliances, foreign direct investment, acquisitions, mergers, links to suppliers and buyers, and so on are important (if not more important) in determining corporate knowledge and profitability.

Examining the individual variables provides some interesting results. Firms with more directors were more likely to possess higher profits. The standardized regression coefficient (β) was as high as 0.489 in Model 3. The standardized regression coefficient results were slightly less for number of directors interlocked, with a high of 0.535 in Model 2 and Model 6. This makes sense as more directors who are able to obtain links to other companies should have a superior grasp of the external business environment.

To assess Saxenian's concept of brain circulation, Canadian firm profitability was tested against international interlocks in addition to those possessing directors with an international education. First, regression results reveal that Canadian firms linked to international firms through an interlocked director experienced higher profitability. The standardized regression coefficient results for international interlocks ranged from 0.24 (Model 3) to 0.570 (Model 1). These results make sense as firms possessing directors that are able to obtain links to international companies should have a superior grasp of the global business environment. The lower coefficients could be explained by the fact that while international knowledge is important, knowledge of the Canadian business climate is more important.

A final variable tested was firms possessing directors with an international education. Here, results were not statistically significant. This result agrees with those presented in the *t*-tests noted earlier. Canadian firms that possessed directors linked to an international company with an international education did not necessarily translate into greater profits. These results do not necessarily agree with this study's hypothesis. Attending a university in a foreign country and being interlocked to a company in that country should provide the greatest amount of knowledge, translating into the greatest profits. Earlier we suggested that if interlocked internationally, 80% of undergraduate alumni are part of an interlock to a firm in the same country that they attended university. But regression results do not suggest that this translates into greater profit.

Table 5b replicates the previous table with four replacements: "# of directors with international education" is supplanted by "international education (excluding USA)"; "# of directors with international education and international interlock" is replaced by "# of directors with international education (excluding USA) and international interlock"; "% of directors with international education" is replaced by "% of directors with international education (excluding USA) and % of directors with international education"; and

Table 5. Regression results for gross profit.

(a)

Covariates	Model 1 b	Model 1 SE	Model 1 β	Model 2 b	Model 2 SE	Model 2 β	Model 3 b	Model 3 SE	Model 3 β	Model 4 b	Model 4 SE	Model 4 β
# of directors							131.02	23.04	0.489*	195.89	35.04	0.332
# of directors interlocked				235.86	48.65	0.535*						
# of international interlocks	885.5	214.44	0.570*	406.37	235.86	0.424*	357.08	226.7	0.409*			
# of directors with international ed.										−228.07	126.6	0.172**
# of directors with international ed. and international interlock												
% of international interlocks	31.52	13.06	0.458*	15.75	13.26	0.379**	8	13.25	0.24**			
% of directors with international ed.										13.24	9.17	0.103
% of directors with international ed. and international interlock												
Intercept		551.82			−54.27			−327.87			−614.76	
R^2		0.34			0.38			0.41		0.41		

Covariates	Model 5 b	Model 5 SE	Model 5 β	Model 6 b	Model 6 SE	Model 6 β	Model 7 b	Model 7 SE	Model 7 β	Model 8 b	Model 8 SE	Model 8 β
# of directors	130.93	.23.07	0.488*				130.93	23.06	0.488*	165.81	28.07	0.265*
# of directors interlocked				216.06	48.71	0.535*						
# of international interlocks	358.04	227.03	0.409*	407.46	236.17	0.424*	358.03	227.03	0.409*			
# of directors with international ed.										−88.01	82.03	−0.166
# of directors with international ed. and international interlock				−661.01	2681.4	−0.012	−604.59	2643.58	−0.01			
% of international interlocks	7.96	13.27	0.34**	15.7	13.27	0.379**	7.95	13.27	0.30***	6.85	9.26	0.134
% of directors with international ed.												
% of directors with international ed. and international interlock	−36.27	158.61	−0.011									
Intercept		−52.8			−52.8			−326.42			−354.019	
R^2		0.38			0.38			0.41			0.41	

(*Continued*)

Table 5. (*Continued*).

(b)

Covariates	Model 4 b	Model 4 SE	Model 4 β	Model 5 b	Model 5 SE	Model 5 β	Model 6 b	Model 6 SE	Model 6 β
# of directors	163.260	20.921	0.360*	145.128	20.963	0.319*			
# of directors interlocked							230.349	46.927	0.251*
# of international interlocks				41.114	154.840	0.013	96.123	188.132	0.229*
# of directors with international ed. (exc USA)	−610.475	211.482	0.188**						
# of directors with international ed. (exc USA) and international interlock							−601.496	2680.979	−0.011
% of international interlocks				−1.882	5.678	−0.015	822.370	610.186	0.073
% of directors with international ed. (exc USA)	63.846	12.747	0.326*						
% of directors with international ed. and international interlock (exc USA)				184.713	28.622	0.293*			
Intercept		−571.7			−422.74			−105.64	
R^2		0.26			0.39			0.38	

Covariates	Model 7 b	Model 7 SE	Model 7 β	Model 8 b	Model 8 SE	Model 8 β	Model 9 b	Model 9 SE	Model 9 β
# of directors	134.615	21.888	0.297*	138.731	21.262	0.306*			
# of directors interlocked							238.321	44.245	0.259*
# of international interlocks	149.600	179.218	0.046						
# of directors with international ed. (exc USA)				963.706	569.822	0.385**	951.664	580.869	.384
# of directors with international ed. (exc USA) and international interlock	−522.860	2640.085	−0.009						
% of international interlocks	794.198	600.914	0.070	51.580	161.911	.016	10.218	164.960	.003
% of directors with international ed. (exc USA)									
% of directors with international ed. and international interlock (exc USA)									
Intercept		−354.41			−347.29			−105.29	
R^2		0.31			0.31			0.38	

"international interlock" is replaced with "% of directors with international education (excluding USA) and international interlock." Results varied a little with total number of directors remaining an important variable while the number of directors interlocked decreased somewhat. Surprisingly, the greatest change was the decreasing importance of international interlocks with the new models. What was of increasing importance in the new models was the % of directors with international education (excluding USA) and % of directors with international education and international interlock (excluding USA).

Conclusion

This study moves beyond the classic unit of analysis in economic geography, the firm as a whole, to concentrate on a specific segment of the business, the firm's board of directors. Consequently, it extends on the work of James O. Wheeler's headquarters location research (Holloway & Wheeler, 1991; Wheeler, 1985, 1986, 1988a, 1988b, 1990), as well as his information flows research (Mitchelson & Wheeler, 1994; Wheeler, 1999; Wheeler & Mitchelson, 1989). By observing the headquarters locations, this study mapped out the embeddedness of cities within the Canadian corporate hierarchy network. Results generally agree with previous research's findings (Rice & Semple, 1993).

Outcomes within this article, associated with Wheeler's research, are then connected to research currently being conducted by economic geographers and sociologists. Findings are placed within the dialogue that examines the relationship between geography and knowledge in today's global economy, especially as it relates to interlocking directorates (Kentor & Jang, 2004, 2004; O'Hagan and Green, 2002a, 2002b; van Veen & Kratzer, 2011) and brain circulation (Jöns, 2009; Kuznetsov, 2006; Saxenian, 2002, 2005, 2006). As highlighted by McCann and Ward (2011, p. xvii), cities are important nodes in a globalizing world and "focusing on the practices that constitute cities as sites of intersection between network typologies and territorial legacies is an important area of study." This research was accomplished through an examination of the spatial relationship between where directors attended universities and the interlocks that they are a part of. Emphasis is placed on cross-border relationships to determine if a correlation exists between Canadian firms linked to international corporations via a director with an international education. We argue that exposure to international institutions and international businesses linkages should result in better corporate performance in today's global world.

If interlocked internationally, a remarkable 80% of alumni are part of an interlock in the same country where they attended university. This finding supports Saxenian's theory. When interlocks to American companies were omitted, the analogous association decreased significantly, revealing the importance of American alumni to Canadian companies and the firms to which they are interlocked. However, a substantial relationship was found between alumni that are part of an interlock in the same country that they attended university, even outside the United States. Geographically, previous educational experience in a foreign market makes directors desirable to Canadian companies. Findings also revealed a relationship between size of firm and the acquiring of international directors. Low-revenue firms are more likely to acquire directors with personal histories that are geographically closer, while larger firms are more likely to acquire directors with international personal histories

It is important to note, however, *t*-test and regression results revealed that a Canadian firm interlocked to an international firm with a director with international education did not necessarily translate into greater profits when American alumni were included.

However, directors with an international education, excluding American alumni, were found to have a positive and significant impact on corporate profits. Just being linked to an international company through a common director did translate into more profitability for Canadian firms. With an increased understanding of global companies, this would suggest that Canadian companies make better business decisions that translate into greater profits.

Results from this study provide the foundation for further research to specifically examine the subset of directors who possess an education background and interlock contained within the same country. Are companies interlocked through a director better linked to the Canadian economy (and vice versa), such as through suppliers and buyers? If an international interlock is broken, how many secondary links are broken from an international network? This can be explored through the concept of betweenness with a truly international data set. Such research would further test if these individuals act as brokers and ultimately provide evidence to address an area of research with important ramifications for knowledge transfer and firm competitiveness.

Disclosure statement

No potential conflict of interest was reported by the authors.

References

Amin, Ash (2004). An institutionalist perspective on regional economic development. In Trevor Barnes, Jamie Peck, Eric Sheppard, & Adam Tickell (Eds.), *Reading economic geography* (pp. 48–58). Victoria, Australia: Blackwell Publishing.

Amin, Ash, & Cohendet, Patrick (1999). Learning and adaptation in decentralised business networks. *Environment and Planning D: Society and Space*, *17*, 87–104.

Bathelt, Harold, Malmberg, Anders, & Maskell, Peter (2004). Clusters and knowledge: Local buzz, global pipelines and the process of knowledge creation. *Progress in Human Geography*, *28*, 31–56.

Baysinger, Barry, & Butler, Henry (1985). Corporate governance and the board of directors: Performance effects of changes in board composition. *Journal of Law Economics and Organization*, *1*, 101–124.

Burt, Ronald (1983). *Corporate profits and co-optation*. New York, NY: Alliance.

Carrington, Peter J. (1981). *Horizontal co-optation through corporate interlocks* (PhD thesis). Dept. Sociology, University of Toronto, Toronto.

Castells, Manuel (1996). *The rise of the network society*. Oxford: Blackwell.

Chirico, Francesco, & Salvato, Carlo (2008). Knowledge integration and dynamic organizational adaptation in family firms. *Family Business Review*, *21*(2), 169–181.

Connelly, Brian, Johnson, Jon, Tihanyi, Laszlo, & Ellstrand, Allen (2010). More than adopters: Competing influences in the interlocking directorate. *Organization Science*, *22*, 688–703.

Cumbers, Andrew, & MacKinnon, Danny (2004). Introduction: Clusters in urban and regional development. *Urban Studies*, *41*, 959–969.

Drucker, Peter F. (2005). *The effective executive*. New York, NY: Harper.

Fligstein, Neil, & Brantley, Peter (1992). Bank control, owner control, or organization dynamics. *American Journal of Sociology*, *98*, 280–307.

Folta, Timothy, Cooper, Arnold, & Baik, Yoon-suk (2006). Geographic cluster size and firm performance. *Journal of Business Venturing*, *21*, 217–242.

Granovetter, Mark (2005). The impact of social structure on economic outcomes. *The Journal of Economic Perspectives*, *19*, 33–50.

Green, Milford (1980). *A geographical examination of interlocking directorates for large American corporations* (Unpublished PhD dissertation). Department of Geography, the Ohio State University: Columbus, OH.

Green, Milford, & Semple, Ronald (1981). The corporate interlocking directorate as an urban spatial information network. *Urban Geography*, *2*, 148–160.

Hassink, Robert (2005). How to unlock regional economies from path dependency? From learning region to learning cluster. *European Planning Studies*, *13*, 521–535.

Holloway, Steven, & Wheeler, James (1991). Corporate headquarters relocation and changes in metropolitan corporate dominance, 1980–1987. *Economic Geography*, *67*, 54–74.

Jones, Andrew (2008). Beyond embeddedness: Economic practices and the invisible dimensions of transnational business activity. *Progress in Human Geography*, *32*, 71–88.

Jöns, Heike (2009). 'Brain circulation' and transnational knowledge networks: Studying long-term effects of academic mobility to Germany, 1954–2000. *Global Networks*, *9*, 315–338.

Kentor, Jeffrey, & Jang, Yong Suk (2004). Yes, there is a (growing) transnational business community: A study of global interlocking directorates 1983–98. *International Sociology*, *19*, 355–368.

Kono, Clifford, Palmer, Donald, Friedland, Roger, & Zafonte, Matthew (1998). Lost in space: The geography of corporate interlocking directorates. *American Journal of Sociology*, *103*, 863–911.

Kuznetsov, Yevgeny (2006). *Diaspora networks and international migration of skills: How countries can draw on their talent abroad*. Washington, DC: WBI Development Studies.

Kuznetsov, Yevgeny, & Sabel, Charles (2006). International migration of talent, diaspora networks, and development: Overview of main issues. In Yevgeny Kuznetsov (Ed.), *Diaspora networks and the international migration of skills: How countries can draw on their talent abroad* (pp. 3–20). Washington, DC: World Bank.

Malecki, Edward (2010a). On diamonds, clusters, and regional development. In Robert Huggins & Hiro Izushi (Eds.), *Competition, competitive advantage, and clusters*. Oxford: Oxford University Press.

Malecki, Edward (2010b). Global knowledge and creativity: New challenges for firms and regions. *Regional Studies*, *44*, 1033–1052.

Malmberg, Anders, & Maskell, Peter (2010). An evolutionary approach to localized learning and spatial clustering. In Ron Boschma & Ron Martin (Eds.), *Handbook of evolutionary economic geography*. Oxford: Edward Elgar.

Malmberg, Anders, Solvell, Orjan, & Zander, Ivo (1996). Spatial clustering, local accumulation of knowledge and firm competitiveness. *Geografiska Annaler*, *78*, 85–97.

Marshall, Alfred (1920). *Principles of economics*. New York, NY: Macmillan.

Maskell, Peter (2009). Towards a knowledge-based theory of the geographical cluster. In Ron Martin (Ed.), *The economy: Critical essays in human geography* (Contemporary Foundations of Space and Place vol. 8). London: Ashgate.

Maskell, Peter, & Malmberg, Anders (2007). Myopia, knowledge development and cluster evolution. *Journal of Economic Geography*, *7*, 603–618.

McCann, Eugene, & Ward, Kevin (Eds.). (2011). Introduction. Urban assemblage: Territories, relations, practices, and power. In *Mobile urbanism: Cities and policymaking in the global age* (pp. xiiii–xxxv). Minneapolis: University of Minnesota Press.

Mitchelson, Ronald, & Wheeler, James (1994). The flow of information in a global economy: The role of the American urban system in 1990. *Annals of the Association of American Geographers*, *84*, 87–107.

Morgan, Kevin (2007). The learning region: Institutions, innovation and regional renewal. *Regional Studies*, *41*, S147–S159.

Nonaka, Ikujiro, & Takeuchi, Hirotaka (1995). *The knowledge-creating company: How Japanese companies create the dynamics of innovation*. New York, NY: Oxford.

O'Hagan, Sean, & Green, Milford (2002a). Tacit knowledge transfer via interlocking directorates: A comparison of Canada and the United States. *Geografiska Annaler, Series B: Human Geography*, *84*, 49–63.

O'Hagan, Sean, & Green, Milford (2002b). Interlocking directorates: An example of tacit knowledge transfer. *Urban Geography*, *23*, 154–179.

O'Hagan, Sean, & Green, Milford (2004). Corporate knowledge transfer via interlocking directorates: A network analysis approach. *Geoforum*, *35*, 127–139.

O'Hagan, Sean, & Rice, Murray D. (2012). The geography of corporate directors: Personal backgrounds, firm and regional success. *Professional Geographer*, *64*, 586–601.

Pennings, Johannes (1980). *Interlocking directorates*. San Francisco, CA: Jossey-Bass.

Pfeffer, Jeffrey, & Salancik, Gerald (1978). *The external control of organizations: A resource dependence perspective*. New York, NY: Harper & Row.

Rice, Murray (2006). Dominant national centres: A comparative analysis of the headquarters communities of New York and Toronto. *Industrial Geographer*, *4*, 28–50.

Rice, Murray, & Semple, Keith (1993). Spatial interlocking directorates in the Canadian urban system, 1971–1989. *Urban Geography, 14*, 375–396.

Rice, Murray D., Tierney, Sean, O'Hagan, Sean, Lyons, Donald, & Green, Milford (2012). Knowledge, influence, and firm-level change: A geographic analysis of board membership associated with growing and declining businesses in the Canadian economy. *Geoforum, 43*, 959–968.

Robinson, Jennifer (2011). The spaces of circulating knowledge: City strategies and global urban governmentality. In Eugene McCann & Kevin Ward (Eds.), *Mobile urbanism: Cities and policymaking in the global age* (pp. 15–40). Minneapolis: University of Minnesota Press.

Saxenian, AnnaLee (2002). Brain circulation: How high-skill immigration makes everyone better off. *The Brookings Review, 20*, 28–31.

Saxenian, AnnaLee (2005). From brain drain to brain circulation: Transnational communities and regional upgrading in India and China. *Studies in Comparative International Development, 40*, 35–61.

Saxenian, AnnaLee (2006). *The new Argonauts: Regional advantage in a global economy.* Cambridge, MA: Harvard University Press.

Singh, Val (2007). Ethnic diversity on top corporate boards: A resource dependency perspective. *The International Journal of Human Resource Management, 18*, 2128–2146.

Sklair, Leslie (2001). *The transnational capitalist class*. Oxford: Blackwell.

Statistics Canada. (2006). *Community profiles from the 2006 census, Statistics Canada – Census subdivision.* Ottawa: Government of Canada.

Storper, Michael (1997). *The regional world: Territorial development in a global economy.* New York, NY: Guilford Press.

Strauss, Erik, Nevries, Pascal, & Weber, Juergen (2010). *The emergence of management control systems: A resource dependency perspective.* Retrieved June 6, 2012, from http://papers.ssrn.com/sol3/papers.cfm?abstract_id=1881464

Ter Wal, Anne & Boschma, Ron (2009). Applying social network analysis in economic geography: Framing some key analytic issues. *The Annals of Regional Science, 43*, 739–756.

Vallance, Paul (2007). Rethinking economic geographies of knowledge. *Geography Compass, 1*, 797–813.

van Veen, Kees, & Kratzer, Jan (2011). National and international interlocking directorates within Europe: Corporate networks within and among fifteen European countries. *Economy and Society, 40*, 1–25.

Wheeler, James (1985). The U.S. metropolitan corporate and population hierarchies, 1960–1980. *Geografiska Annaler B, 67*, 89–97.

Wheeler, James (1986). Corporate spatial links with financial institutions: The role of the metropolitan hierarchy. *Annals of the Association of American Geographers, 76*, 262–274.

Wheeler, James (1988a). The corporate role of large metropolitan areas in the United States. *Growth and Change, 19*, 75–86.

Wheeler, James (1988b). Spatial ownership links of major corporations: The Dallas and Pittsburgh examples. *Economic Geography, 64*, 1–16.

Wheeler, James (1990). Corporate role of New York City in the metropolitan hierarchy. *Geographical Review, 80*, 370–381.

Wheeler, James (1999). Local information links to the national metropolitan hierarchy: The southeastern United States. *Environment and Planning A, 31*, 841–854.

Wheeler, James, & Mitchelson, Ronald (1989). Information flows among major metropolitan areas in the United States. *Annals of the Association of American Geographers, 79*, 523–543.

Zelinsky, Wilbur (1971). The hypothesis of the mobility transition. *Geographical Review, 61*, 219–249.

Fast-growing firms as elements of change in Canada's headquarters city system

Murray D. Rice[a], Donald I. Lyons[b] and Sean B. O'Hagan[c]

[a]*Department of Geography, University of North Texas, Denton, USA;* [b]*Department of Geography, University College Cork, Cork, Ireland;* [c]*Department of Geography, Nipissing University, North Bay, Canada*

This research explores the factors that shape the evolving geographic distribution of business headquarters (HQ) activity. We address an understudied influence on HQ geographies: metropolitan HQ changes driven by the process of small, rapidly expanding businesses growing into mature companies. This investigation focuses on the developmental paths followed by fast-growing firms (FGFs) and the geographic distinctions that can be observed in a FGF tracking study of Canada's metropolitan regions from 1987 to 2005. Our research findings indicate that geography plays an important role in this development, as FGF tracking records throughout Canada's metropolitan areas diverge sharply. We find that most FGFs that emerged in Vancouver and Toronto continued as ongoing businesses following their rapid growth phase, while a high proportion of FGFs based in Montréal and Calgary did not. These results contribute to a greater understanding of metropolitan economies, business development, and HQ location in Canada

Introduction

In times of economic instability, business activities that provide long-term, high-paying jobs are highly valued by both the public and private sectors. Business headquarters (HQ) activities fit this profile well. Communities that possess corporate HQ tend to place a high priority on retaining the HQ they already host and on efforts to attract new HQ. Examples of municipalities seeking to attract HQ activity can be found across North America. The regional repercussions of HQ relocation by Boeing (from Seattle to Chicago), AT&T (from New York to San Antonio and then San Antonio to Dallas), and Imperial Oil (from Toronto to Calgary) underline the significance of interurban transfers of corporate power, especially among the most elite of corporations. Governments at multiple levels often play a key role in influencing HQ location decisions. For example, in 2002 the City Council of Fort Worth, Texas, approved up to $96 million in tax incentives in an effort to help the electronics retailer, RadioShack, establish a new HQ campus and stay in the city. Furthermore, in the midst of the 2008–2010 recession, Fort Worth's city government provided a new round of $10.7 million in tax rebates to again keep RadioShack from pursuing possible relocation options (Lee, 2010). Such initiatives are not rare (Jean, 2013). HQ are widely viewed as key regional assets and generators of economic activity (Card, Hallock, & Moretti, 2010).

Any regional effort to gain HQ activity must be rooted in a deep understanding of how corporations and HQ grow and change. Elite corporate activities, including HQ location decisions, exist within a complex matrix of political, social, and economic relationships; these relationships have an important impact on the places where businesses choose to locate their HQ (Knoben, 2011). However, the factors that shape the trends leading to the emergence and decline of regional HQ centers have been defined only in a most general sense (Rice & Lyons, 2010). Furthermore, much of the existing HQ research focuses on large firms, such as the members of the *Fortune 500* group of companies in the United States and the *Financial Post 500* group of companies in Canada (Gong & Wheeler, 2007; Holloway & Wheeler, 1991; Semple, 1996). Large firms do have substantial economic impacts, based on the size of their operations and their workforce numbers; however, such companies are by no means the only relevant subjects for research if the goal is a deep understanding of HQ growth and locational dynamics. A large-firm research focus is inadequate on its own, in part simply because even the most influential businesses do not begin as large firms. How firms emerge as major economic players, and thus grow to become key members of a regional HQ complex, is an important but neglected part of this scholarly discussion. Moreover, in the process of firm growth, the corporate footprint of businesses expands, meaning that firms' economic impact diffuses and encompasses warehouses, stores, factories, and regional offices in an increasing number of metropolitan markets. Thus, how firms develop (and where decision-makers and other operations are based as firms emerge) is an important but complex question that relates directly to the core debates of development and business research in economic geography.

It is from this perspective that our study joins a growing body of research in considering a broad range of business activities that give insight into the complex phenomenon of firm development and location. In line with the preceding discussion, we assert that HQ have an important and direct connection to regional prosperity and that understanding the factors that contribute to HQ expansion is a crucial part of research on business location and development. In a general, business development context, it is important to note that fast-growing firms (FGFs) have been the subject of extensive inquiry over the past three decades, beginning with Birch's (1981, 1987) seminal investigations into small firms and economic growth.[1] In subsequent years, substantial research has examined these dynamic firms and their characteristics and impacts (Henrekson & Johansson, 2010; Markman & Gartner, 2002; Stam, 2005). Geographers interested in HQ location have identified rapid firm growth as an important complement to large-firm HQ research (Rice & Lyons, 2010; Wheeler, 1990), and it is this particular HQ/rapid growth connection that we pursue in this article.

The exploratory research we outline in this article addresses a Canadian case study to give insight into the extended development track associated with FGFs. While FGF development relates to geographically complex considerations (i.e., the locations of warehouses, stores, factories, and regional offices), our study highlights HQ location specifically. A trio of time-, space-, and industry-based questions provides the focus for the analysis: *temporally*, what happens to Canadian FGFs in the years after their peak growth; *spatially*, which Canadian metropolitan regions are most successful in converting the necessarily transitory nature of FGF activity (and associated HQ) into a long-term corporate presence; and, finally, in terms of the *metropolitan industrial context* for FGF development, what connections might be extracted between the industrial composition of a regional community of FGFs and the metropolitan business development outcomes we observe? The following section provides some brief background on the HQ/firm growth nexus and outlines the above questions and our Canadian case study in more detail.

Research context in the literature of geography and business

The foundation for this research is the wide-ranging literature on the geography of corporate decision-making and high-level business analysis ("quaternary sector") activities that collectively manage and administer the economy as a whole (Gong & Wheeler, 2007; Holloway & Wheeler, 1991; Rice & Lyons, 2010; Rice, Tierney, O'Hagan, Lyons, & Green, 2012; Semple, 1985). Beginning with an interest in the largest firms and corporate centers (Goodwin, 1965; Quante, 1976), quaternary location research has documented the changes that have characterized the North American HQ system as firms have dispersed away from established centers such as New York, Chicago, Toronto, and Montréal, and toward emerging major regional centers such as Atlanta, Dallas, Calgary, and Vancouver (Gong & Wheeler, 2007; Semple, 1996; Stephens & Holly, 1981). Quaternary research since the 1980s has broadened the focus of HQ location research to encompass a host of factors related to HQ activity, including regional and subsidiary offices (Meyer & Green, 2003; Rice & Pooler, 2009; Wheeler, 1988a), the geography of high-level information exchange (Mitchelson & Wheeler, 1994; Walcott & Wheeler, 2001), director locations and linkages (Green & Semple, 1981; O'Hagan, Rice, & Green, 2008; Rice et al., 2012), and the geography of ownership and investment (Green, 1993, 2004; Wheeler, 1988b). The diversification of quaternary location research has improved scholarly understandings of the multitude of factors that influence corporate HQ locations and relocations. This broadened perspective makes the development of a research link between HQ location in general and FGF activity in particular a logical next step.

One situation in which FGF research is central to an improved understanding of HQ location is in the examination of emerging corporate centers, focusing on how the growth of new industries creates new geographies of elite corporate activity in a national city system (Beyers, 2002). Several North American examples can be cited:

(1) Detroit, Michigan, in the late 1800s was host to many minor, regionally important industrial firms, but by 1950 it had gained an elite status as home to the HQ of the world's most important automobile manufacturers (Poremba, 2003; Woodford, 2001). Detroit is known today as a focal point of automotive design and manufacturing, but it is also important (even with the region's contemporary economic troubles) to acknowledge Detroit's role as host to the key financial and strategic *decision-makers* for North America's automotive giants.

(2) Charlotte, North Carolina, was a regional financial center until the 1990s, but the deregulation of interstate banking, aggressive management, and a series of strategic acquisitions enabled Charlotte's banking community to climb to the top tier of the United States financial sector as a whole by the 2000s (Graves & Kozar, 2010; Lord, 1992). Even after the 2008–2010 recession, Charlotte has continued as a banking center of national stature.

(3) Ottawa provides a Canadian example of a regional corporate emergence. Until the 1970s, the economy of the Ottawa metropolitan region was dominated by its public-sector functions. However, beginning in the 1980s, Canada's national capital emerged as an influential corporate center in the Canadian high-technology sector (Chamberlin & de la Mothe, 2004).

How these and other regions continue to evolve from a position of business obscurity to prominence, and thus play a role in shifting the overall corporate landscape (especially related to emerging economic sectors), is a question that must be addressed from a perspective that explicitly incorporates new firm growth and development alongside an analysis of existing large firms.

Indeed, research that tracks the development of FGFs lies at the core of this area of research. It is important to establish the characteristics of FGF development to ascertain the value of FGFs as economic development mechanisms, and to begin to seek out the characteristics of firms and regions that are connected to the most robust FGF developmental outcomes. Construction of such a corporation-place connection is essential, because location is central to firm competitiveness. As argued by Porter (2000, p. 253), "competitive success is anything but widely and evenly dispersed. Some nations, regions within nations, and cities contain far more competitive firms than others, even at locations nearby."

We acknowledge that firm tracking research in general, and FGF tracking in particular, is not new. Papers focused on the United Kingdom (Parker, Storey, & van Witteloostuijn, 2010), Germany (Almus, 2002), Sweden (Delmar, Davidsson, & Gartner, 2003), Japan (Honjo, 2004), and the United States (Acs, Parsons, & Tracy, 2008) have all addressed the developmental trajectories of FGFs. Among the most important findings of these studies are a high degree of inconsistency in the growth of these firms (Parker et al., 2010), a high degree of sectoral diversity among FGFs, including both technology-intensive and technologically unsophisticated firms (Bishop, Mason, & Robinson, 2009), and a low degree of firm growth explanation attributable to any one variable (Coad, 2009). Thus, the overwhelming consensus of the existing literature is that the FGF phenomenon is highly complex. The multifaceted nature of FGFs is such that continued research into diverse growth factors and regional business settings is required if a comprehensive understanding of FGF activity is to be obtained (Stam, 2010).

The preceding summary shows that many aspects of business establishment and operation have been incorporated into FGF tracking studies. Two further specifics emerge from this discussion. First, the combination of *temporal tracking* and *geographic analysis of FGFs* has not been well developed. As argued above, this particular combination has much potential to contribute to both the FGF and HQ location literatures. Second, although a small body of research has addressed firm growth and development in Canada (Baldwin, Bian, Dupuy, & Gellatly, 2000; Britton & Gilmour, 1978; Rice, 2004; Rice & Lyons, 2007; Steed, 1982), Canada is one national setting that has not been the subject of FGF tracking research. A baseline study is needed. With these two opportunities for contribution in mind, the remainder of this article is devoted to an exploratory tracking analysis of FGF activity over time and by Canadian metropolitan area, with the goal of contributing to the body of research within economic geography on FGFs both in general and specifically as a factor in emerging HQ geographies.

Canadian case study

Research questions

As noted in the previous section, this article presents a spatial–temporal–sectoral analysis of Canada's national network of FGFs, with an examination of FGF activity in three complementary contexts: temporal, geographic, and industrial. The key question for initial research focuses on a profile of the *temporal development paths* followed by these Canadian businesses. What happens to FGFs in the years after these firms have achieved their peak growth? Our study identifies five possible tracks for FGF development:

(1) *Firm continuity*: FGFs that continue in business independently, without merger or acquisition, in the same metropolitan area as their initial location.
(2) *Relocation*: FGFs that continue in business independently, but in a different metropolitan area (in Canada or elsewhere) compared to their initial location.
(3) *Acquired/active*: FGFs purchased by or merged into another firm, but with the original firm maintaining some identifiable business presence (e.g., continuation of the original business name or continued use of original facilities) in the original metropolitan area.
(4) *Acquired/no records*: FGFs purchased or merged into a new firm, with no readily identifiable business presence in their original metropolitan area.
(5) *Defunct/no records*: FGFs for which no record exists of a purchase or merger, and no record exists of any ongoing firm operations (including corporate telephone listings or Web sites); in this case, the firm is likely bankrupt or has otherwise gone out of business.

Each of these five possibilities is mutually exclusive and firms can fall in only one category at a time. However, it is possible for a firm to change status over time; for example, a business falling in the "firm continuity" class in 1 year may fall in the "relocation" class at a later time. Given that no previous research has investigated the extended developmental track of FGFs in a Canadian context, one contribution of this investigation is to provide evidence of the changing status of Canada's most rapidly expanding companies.

Since business development occurs not only over time, but also within a geographic context, a second area of investigation connects our temporal analysis to location by studying the *geography of firm development*. We ask which regions of the country have experienced the highest rates of FGF continuity, and have been able to 'keep' the dynamic firms they generate without losing them through acquisition, bankruptcy, or relocation. Do metropolitan areas located in Canada's most productive economic regions (such as Alberta and British Columbia currently, and Ontario historically) have an advantage in retaining their firms, or do selected cities in other regions also keep their most dynamic firms as these firms grow? Such locally retained firms might be classified as "successes" from a regional economic development perspective.[2]

A final area of inquiry is the role of FGFs in changing Canada's national HQ landscape given the regional *industrial context* for firm development. Past studies have provided evidence to indicate the composition of regional HQ communities broadly (Meyer & Green, 2003; Rice, 2006; Semple, 1996), but these studies have examined large firms exclusively. The literature has yet to place a specific focus on regional differences in the industrial orientation of FGFs. It is important to mention that a small number of studies have established that Canadian FGF activity at the national level is concentrated in particular sectors such as business services, electronics, and biotechnology (Rice, 2004; Rice & Lyons, 2007). However, in the absence of detailed metropolitan analysis, it is impossible to gauge the extent to which the profile of FGF activity at the metropolitan level differs from one urban region to another, and whether these differences might have some relationship with the regional outcomes addressed in our second research question. Thus, the third research question asks how regional centers in Canada's urban system compare in terms of the industrial composition of their FGF communities. To lay out the basis for this exploration, we note two key characteristics of the Canadian HQ city system:

(1) First, very few Canadian metropolitan areas have a HQ community of sufficient size and complexity to support a detailed, regional analysis. In recent decades, Toronto, Montréal, Calgary, and Vancouver have hosted more than 70% of Canada's large parent HQ and in excess of 80% of large subsidiary HQ, while the proportion of Canadian FGFs hosted by these same cities is similarly high (Rice & Lyons, 2007; Rice & Pooler, 2009; Semple, 1996). Only Canada's "Big Four" corporate centers have the number of firms necessary for a detailed industrial analysis.
(2) Second, despite this apparent stability, some important changes should be noted. Of the "Big Four" cities, Calgary and Vancouver have grown dramatically as large-firm HQ hosts in recent years (Rice & Pooler, 2009). Table 1 demonstrates the extent of HQ growth in these two cities in particular. By a large margin, Calgary and Vancouver lead all other Canadian metropolitan regions in HQ growth.

Thus, it is the *rapid emergence* of these specific metropolitan areas as corporate centers and the *size of their regional business communities*, which provide a unique opportunity to investigate how new centers emerge within Canada's national HQ city system. We ask whether FGF analysis can provide some insights into such a metropolitan corporate emergence. Given this overall focus, we confine our scope for the third research question(s) to a detailed analysis of FGF industries and developmental outcomes associated with the dynamic corporate communities of Calgary and Vancouver.

Table 1. Top 10 Canadian metropolitan areas as ranked by large-firm headquarters (HQ) growth, 1986–2006.

HQ growth rank	Metropolitan area	Number of HQs, 1986*	Number of HQs, 2006*	HQ growth, 1986–2006**
1	*Calgary, AB*	*69*	*130*	*+61*
2	*Vancouver, BC*	*66*	*103*	*+37*
3	Quebec City, QC	8	22	+14
4	Regina, SK	9	15	+6
5	Kitchener–Waterloo–Cambridge, ON	10	16	+6
6	Ottawa, ON	21	26	+5
7	Hamilton, ON	18	20	+2
8	Halifax, NS	8	10	+2
9	Guelph, ON	5	7	+2
10	Sherbrooke, QC	1	3	+2

Source: Compiled by the authors from Dun and Bradstreet (1987–2007) and Financial Post (1987, 2007).
Notes: *From the top 1,000 companies in Canada as ranked by annual revenues for 1986 and 2006.
**The major corporate centers of Montreal, Toronto, and Winnipeg all experienced a decline in the number of HQs each hosted from 1986 to 2006 (total metropolitan losses of 47, 38, and 10 firms, respectively).

Data

We examine two primary databases focusing on Canada's national FGF community. The first database consists of the top firms listed in the Canadian business magazine Profit's annual tabulation of Canada's FGFs as ranked by 5-year revenue growth. Beginning with a list published in 1988, based on 1987 financial results, Profit ranked the most rapidly expanding companies of all ownership types, including public, private, cooperative, and government-owned ("Crown") corporations. The initial list for 1987 included 50 firms in total, although Profit expanded this number in subsequent years (up to 200 firms were tabulated annually from 1998 until the present). However, to maintain consistency and to avoid bias by weighting later study years more heavily than earlier ones, we extract only the top 50 firms from each of eight study years: data for 1987 (the earliest year available), 1990, 1992, 1994, 1996, 1998, 2000, and 2002, providing a total of 400 firms and a time span of 15 years in the initial database. Fields included in this database include HQ location, industry, revenues, and revenue growth statistics. Table 2 provides a brief indication of the size of firms covered here; while the businesses studied are not exclusively small (10 have 500 or more employees), small firms are by far the most numerous (177 have fewer than 50 employees).

The Profit database provides a good basis for this investigation, as it is the only national ranking of FGFs published annually in Canada. However, it is important to note the limitations of the Profit ranking. Profit assembles its list from nominated businesses, so some worthy businesses may not be included. However, the prestige associated with inclusion on a national rapidly growing business ranking provides substantial incentive for growing businesses to ensure they are nominated (businesses appearing in the Profit ranking, and regions hosting included firms, routinely publicize their inclusion in the list). Also, Profit's use of revenues for its growth measure directs attention to businesses experiencing a dollar-value expansion in their annual business volumes. Businesses that are growing in dimensions other than overall business volumes, such as profitability or number of employees, are thus not the focus of the Profit database. Rankings based on profits or employees are certainly possible, but would explore a different aspect of the Canadian corporate landscape. Our study uses the Profit list to identify the firms (and associated cities) that are gaining a greater presence in the Canadian economy through heightened business activity.

Table 2. Study firms by number of employees*.

Employee range	Number of firms
1,000 and Greater	4
500–999	6
200–499	22
50–199	77
Less than 50	177
No employment data	114
Total	**400**

Source: Compiled by the authors from Profit Magazine (1988–2003).
Notes: *Profit published data for most but not all businesses in their rankings. We were unsuccessful in obtaining further employment information on any of the 114 firms with no employment data. It is likely that most, if not all, of these 114 firms are small employers.

A second database extends on the first, with the purpose of tracking the ongoing status of each of the 400 FGFs included in the initial database. This tracking database specifies the status of these FGFs from the year of their first appearance in the *Profit* listing until 2005, the final year of data inclusion for the tracking portion of our study. Thus, our total tracking analysis occurs over the period 1987–2005. Information for this tracking database comes from a collection of Canadian business directories, including The Blue Book of Canadian Business, Dun & Bradstreet's Canadian Key Business Directory, and the Financial Post publications: FP Corporate Surveys, FP Investor Reports, and FP Predecessor & Defunct. The study also relies on Yellow Pages business listings and corporate web information, gathered through Google, to check for active telephone records and corporate web pages for all FGFs in the study database. Reliance on a wide variety of separate business directories and sources, all of which attempt to be comprehensive, helps to ensure that the data collected for the tracking database are as complete as possible. Using these resources, this project traces the status and HQ locations for each FGF included in the initial database for 2, 5, and 10 years past the year of the company's attainment of FGF status in the Profit list (note, however, that 5- and/or 10-year tracking does not exist for FGFs included in the later years of the database, given the 2005 end date for firm tracking data). Analysis of this tracking database forms the core of this study.

Results

The temporal dimension

First, we track the changing status of FGFs for Canada as a whole. Table 3 summarizes the developmental profile of firms in the Canadian FGF database. In Table 3, the "base year" is the year of each firm's appearance as an FGF, and we track firm status 2, 5, and 10 years following each corporation's appearance among the top 50 business in the Profit annual ranking. Clearly, firm continuity (i.e., continued operation as an independent firm in the same metropolitan area) is the dominant outcome for all time periods. As is to be expected, Table 3 shows a declining proportion of firms maintaining "firm continuity" status over time. However, given the inherently dynamic nature of the FGF group of firms, we consider the observed 10-year firm continuity rate of 56.0% to be robust. The

Table 3. Developmental profile of Canadian FGFs.

Category	Base + 2 years (percentage of all firms)	Base + 5 years (percentage of all firms)	Base + 10 year (percentage of all firms)
Firm continuity	76.3	67.9	56.0
Relocation	2.0	0.8	3.0
Acquired/active	2.9	4.0	6.0
Acquired/no records	3.8	8.4	11.3
Defunct/no records	15.0	18.9	23.8
Total firms tracked*	400	350	200

Source: Compiled by the authors from Dun and Bradstreet (1990–2006), Canadian Newspaper Services International (1990–2006), Financial Post (1990–2006a), Financial Post (1990–2006b), Financial Post (1990–2006c), and online searches for business mentions via Google.

Notes: *The number of firms tracked varies because the end year of the initial FGF database (2002) is less than 10 years from the end year of the tracking database (2005). For example, FGFs in 2002 had neither a 10-year nor a 5-year record to track in the database as collected, but they did have a 2-year record, while firms from 2000 had 2- and 5-year records, but not 10-year records.

second-ranking 10-year outcome is "defunct/no records" (23.8% of FGFs), which highlights the risk associated with consideration of FGFs as a class (an "up or out" situation—collectively, 79.8% of FGFs either continue as independent, mature firms, or die). Despite this, the fact that firm continuity almost doubles firm discontinuation as a 10-year outcome is an indication of the potential promise of FGF activity as a source of growth within the Canadian economy. This result indicates that further research into FGFs in a Canadian context is warranted. Overall, Table 3 provides important evidence pointing to a need for more understanding of the processes related to firm growth.

Beyond this, Table 3 also demonstrates the low importance of relocation as a mechanism of change among the companies in the database (3.0% of all FGFs after 10 years). Indeed, firm acquisition is a more important developmental factor than relocation. Within the acquired firms class overall, "acquired/no records" is a more frequent outcome than "acquired/active", meaning that FGFs are most often folded into the acquiring firm's corporate structure, as opposed to continuing with their previous public identity. However, it is again important to note the low rate of acquisitions relative to firm continuity and firm discontinuation.

The geographic dimension

Our second task is to break out the geographic detail aggregated into the national totals provided in Table 3. The analysis here aims to summarize the varying abilities of the business communities in Canada's major metropolitan regions to nurture and maintain the FGFs they generate. The results here are represented in two tables: Table 4 deals with Canada's "Big Four" corporate centers (Toronto, Montréal, Vancouver, and Calgary), while Table 5 covers the next tier of mid-size business centers. The mid-size metropolitan areas mentioned in Table 5 are defined as a group in this context by the much smaller size of their FGF communities in comparison with Canada's "Big Four." Segmentation of results into these two tables is necessary to achieve a fair comparison of cities and continuity rates; cities with a large number of FGFs are compared directly and exclusively with similar urban centers, while cities with moderate FGF totals are also compared exclusively with other, similar cities. Only cities that have produced five or more FGFs over the entire 1987–2002 initial data period, and have at least one firm with a 10-year record to track, are included in these two tables.

Table 4. Firm continuity rates for FGFs in Canada's "Big Four" metropolitan areas, ranked by 10-year continuity rate.

Metropolitan area	Number of FGFs generated, 1987–2002	Continuing FGFs as a percentage of all FGFs in the city		
		Base + 2 years	Base + 5 years	Base + 10 years
Vancouver, BC	55	89.1	71.8	66.7
Toronto, ON	136	75.0	71.7	62.3
Calgary, AB	34	76.5	60.9	42.9
Montréal, QC	36	58.3	54.8	28.6

Source: Compiled by the authors from Dun and Bradstreet (1990–2006), Canadian Newspaper Services International (1990–2006), Financial Post (1990–2006a), Financial Post (1990–2006b), Financial Post (1990–2006c), and online searches for business mentions via Google.

Table 5. Firm continuity rates for other Canadian metropolitan areas, ranked by 10-year continuity rate.

Metropolitan area	Number of FGFs generated, 1987–2002	Continuing FGFs as a percentage of all FGFs in the city		
		Base + 2 years	Base + 5 years	Base + 10 years
Winnipeg, MB	6	100.0	100.0	100.0
Kitchener–Waterloo–Cambridge, ON	11	87.5	87.5	75.0
Saint John, NB	5	100.0	80.0	75.0
Edmonton, AB	12	66.7	62.5	62.5
Ottawa-Gatineau, ON/QC	19	78.9	62.5	61.5
London, ON	10	80.0	42.9	14.3

Source: Compiled by the authors from Dun and Bradstreet (1990–2006), Canadian Newspaper Services International (1990–2006), Financial Post (1990–2006a), Financial Post (1990–2006b), Financial Post (1990–2006c), and online searches for business mentions via Google.

Table 4 reflects the dynamic nature of the "Big Four," with all four cities generating in excess of 30 FGFs over the study period. Table 4 also shows that the robust metropolitan economies of Vancouver and Toronto both have 5-year firm continuity rates in excess of the national average rate of 67.9%; the two corporate centers also exceed the national 10-year rate of 56.0%. By contrast, despite its rapidly expanding economy and large, multinational corporate base, the otherwise-dynamic metropolitan area of Calgary lags behind the 5- and 10-year national averages, with continuity rates of 60.9% and 42.9%, respectively. Montréal's firm continuity performance trails well behind Calgary, with an extremely low 10-year continuity rate of 28.6%, as well as lower-than-national-average 2- and 5-year rates. These results for Montréal and Calgary are surprising, given the continuing importance of the two cities relative to the Canadian economy overall and the indications of economic dynamism embodied in their FGF generation rates.

These results provide an interesting contrast when considered in relation to other recent analysis of HQ trends for these four cities. For example, while Rice and Pooler's (2009) large-firm analysis agrees with our findings on Vancouver (growing rapidly) and Montréal (declining), their analysis depicts Calgary as leading the country in large-firm HQ growth while Toronto ranks second only to Montréal in large-firm HQ decline. In Montréal's case, perhaps a decline in large-firm HQ has provided opportunity for FGF development. The conclusion of our article further discusses these research contrasts and more considerations related to each of the "Big Four".

Table 5 provides a parallel analysis of contrasting metropolitan fortunes for mid-size corporate centers. Among these metropolitan areas, Winnipeg's performance leads all others, given its 10-year success in maintaining 100% of the FGFs generated in the city. Ottawa, Waterloo, Saint John, and Edmonton all exceeded the national 10-year average continuity rate. Among these mid-size cities, only London performed below the national average, with an extremely low 10-year FGF maintenance rate of 14.3%. On average, the FGF trend for these mid-size centers is remarkably positive and could indicate that one or more of these urban areas are in the early stages of a larger corporate emergence.

Despite this evidence of positive mid-size center performance, the "Big Four" metropolitan areas truly dominate in terms of FGF activity. Because of the sheer size and importance of these metropolitan areas, Table 6 provides additional FGF tracking detail

Table 6. Developmental profile of FGFs for Canada's "Big Four" metropolitan areas (10-year firms and outcomes only)*.

Metropolitan area	Acquired/active (%)	Acquired/no records (%)	Defunct/no records (%)	Relocated (%)	Firm continuity (%)	Total firms in city
Vancouver, BC	0.0	6.7	23.3	3.3	66.7	30
Toronto, ON	5.7	11.3	20.7	0.0	62.3	53
Calgary, AB	0.0	28.5	14.3	14.3	42.8	14
Montréal, QC	14.3	14.3	38.1	4.8	28.5	21
Totals (absolute numbers and %)	6 (5.1)	15 (12.7)	28 (23.7)	4 (3.4)	65 (55.1)	118 (100.0)

Source: Compiled by the authors from Dun and Bradstreet (1990–2006), Canadian Newspaper Services International (1990–2006), Financial Post (1990–2006a), Financial Post (1990–2006b), Financial Post (1990–2006c), and online searches for business mentions via Google.

Note: *Percentages refer to the proportion of firms in the metropolitan area that fall in each category. For example, 6.7% of Vancouver's firms were in the "acquired/no records" category at the 10-year mark.

for the "Big Four." Focusing exclusively on 10-year outcomes, the table breaks down the various firm status possibilities for each city. The findings here emphasize the importance of Vancouver and Toronto as Canadian leaders of FGF development, not just in terms of FGF generation, but also in firm retention and continuing firm development. Vancouver's superior 10-year performance, 66.7% FGF continuity and relocation of only 3.3% of businesses, is remarkable given the city's peripheral location relative to major national and continental markets.[3] Growth-oriented firms, conceived broadly, have substantial incentive to relocate to more locationally central and better connected markets with respect to the Canadian and North American economies, so Vancouver's hold on these firms is impressive.

The analysis for Toronto also provides evidence of economic success. The city's 53 10-year FGFs achieve a 62.3% continuity rate, but just as noteworthy is the complete absence of relocation: no FGF shifted location away from Toronto to any other metropolitan area at any point during the study period. This finding alone is a good indication of the ongoing strength of Toronto's business environment and provides important context to Rice and Pooler's (2009) large-firm decline findings. The fact that Toronto's FGF leaders unanimously view Toronto as the preferred location for their HQ, over a long period of time, is an important sign of confidence in the continued business growth prospects of Canada's largest metropolitan region.

Conversely, Montréal-based FGFs (21 in total) experienced a firm continuity rate of only 28.5%. Additionally, 38.1% of Montréal FGFs fell into the "defunct/no records" classification (compared to 23.8% nationally). Calgary also saw poor performance from its FGF community, with 14.3% of its firms relocating and an additional 14.3% of firms falling into the "acquired/no records" class. Again, relative to the large-firm trends established by Rice and Pooler (2009), the Montréal results are consistent (the large-firm and FGF trends are similarly poor), but the Calgary results diverge (the large-firm results are highly positive, while the FGF result is poor). All of these findings receive further comment in the conclusion.

The industrial dimension

The final goal of our research is to investigate whether an industrial analysis of FGF composition can provide insight into FGF track records by region. These results are notable given the lack of comparative FGF regional analysis in the existing literature, especially with reference to Canada's metropolitan regions. However, given the context provided by the results for the second research question, the analysis associated with this third and final question takes on an added dimension: what might contribute to Vancouver and Calgary, two growing regional corporate centers, differing so widely in terms of FGF tracking outcomes? This section provides analysis related to this issue.

Table 7 provides some perspective on the contrasting outcomes of FGF development in Calgary and Vancouver by analyzing the industrial composition (via four-digit NAICS codes) of FGFs in the two metropolitan regions. While Table 7 shows no causal relationship between industrial composition and FGF development, these results (in combination with the previous tracking results of Table 6) provide a compelling foundation for hypothesis generation and further research. Among the most important insights from Table 7 is that both Calgary and Vancouver have dominant sectors in their FGF communities. However, Calgary's top FGF industry (NAICS sector 2111, Oil and Gas Extraction, 27.2% of metropolitan FGFs) plays a much more substantial metropolitan role than does the top FGF industry for Vancouver (NAICS sector 5413, Architectural and Engineering

Table 7. NAICS sectors for FGFs in Calgary and Vancouver, 1987–2005 tracking analysis.

Metropolitan area	NAICS sector	Number of FGFs	Sector percentage of all metro FGFs
Calgary			
	2111: Oil and Gas Extraction	9	27.3
	5413: Architectural, Engineering, & Related Services	3	9.1
	2131: Support Activities for Mining	2	6.1
	5112: Software Publishers	2	6.1
	17 additional sectors	1 each	3.0 each
	Calgary metropolitan area total	***33***	***100.0***
Vancouver			
	5413: Architectural, Engineering, & Related Services	6	10.3
	3345: Navigational, Measuring, and Control Institution*	4	6.9
	3162: Footwear Manufacturing	3	5.2
	4431: Electronics and Appliance Stores	3	5.2
	5112: Software Publishers*	3	5.2
	5179: Other Telecommunications*	3	5.2
	5415: Computer Systems Design and Related Services*	3	5.2
	3113: Sugar and Confectionery Product Manufacturing**	2	3.4
	3121: Beverage Manufacturing**	2	3.4
	4234: Professional Equipment and Supply Whole	2	3.4
	5182: Data Processing, Hosting, and Related Services*	2	3.4
	5416: Management, Scientific, and Technical Consulting*	2	3.4
	23 additional sectors	1 each	1.7 each
	Vancouver metropolitan area total	***58***	***100.0***

Source: Compiled by the authors from Dun and Bradstreet (1990–2006), Canadian Newspaper Services International (1990–2006), Financial Post (1990–2006a), Financial Post (1990–2006b), Financial Post (1990–2006c), and online searches for business mentions via Google.
Notes: *Activities related to electronic instrumentation and communications.
**Activities related to food manufacturing.

Services, 10.3% of metropolitan FGFs). Thus, Calgary has a large proportion of its FGFs competing directly against each other within a single dominant industry; this provides a plausible explanation for the high rates of business acquisition (and firm discontinuation more broadly) observed in that city. A much smaller proportion of Vancouver FGFs are in direct competition with each other, which is likely related to this city's much higher rates of firm continuity.

Additionally, examination of the NAICS sectors represented in Table 7 indicates a potential for localization economies among several multiple-FGF sectors present in Vancouver. Vancouver's FGFs include a group of high-technology firms involving NAICS sectors 5112, 5179, 5182, 5413, and 5415 and a group of food-related firms involving NAICS sectors 3113 and 3121. Calgary's FGF community, by contrast, is highly concentrated by industry, involving a dominant and overwhelming focus on oil and gas. Although both cities have a diverse group of single-FGF sectors, these results highlight an interesting divergence between the two cities. This distinction becomes even more instructive when considered in conjunction with the previous FGF findings for each city. Can FGF industrial structure be related to FGF development outcomes for the two cities?

At this point, it is useful to acknowledge that FGFs do not exist in isolation from the rest of the economy (regional, national, and global) and that the operation of other business types in a metropolitan area must be considered in conceptualizing the factors contributing to regional FGF development and success. Thus, it is important to bring large firms (as the most prominent group of businesses in the country) into consideration as well. Table 8, based on the Financial Post's *FP 500* ranking of large Canadian firms for 2005 (the final year of the FGF tracking analysis), provides a large-firm summary by industry for Calgary and Vancouver. The difference between the two cities is clear: with 50.6% of its large firms concentrated in its oil and gas sector, Calgary has a far higher level of industrial concentration compared to Vancouver (the premier large-firm sector in Vancouver is metal ore mining with 8.6% of the region's large firms). Yet, even beyond contrasting concentration levels, Table 8 highlights yet another important difference between the cities:

- Calgary's industrially concentrated large-firm community has a high degree of coincidence with the region's FGF community as viewed by NAICS industry type. Both groups feature energy and resource-related firms as the region's premier business group.
- Vancouver's diverse group of FGFs has a lower degree of relationship with the region's diverse group of large firms. For example, Vancouver's two leading large-firm industries (2122: Metal Ore Mining, and 3221: Pulp and Paper) have no counterparts among the top 12 FGF industries cited in the Vancouver portion of Table 6.

Concentration in a limited range of industrial classifications is thus a pervasive characteristic of Calgary's metropolitan corporate community, while Greater Vancouver's corporations are inherently more diverse by economic sector. While this exploratory paper can only note these industrial configurations alongside parallel observations of FGF developmental failure (Calgary) and success (Vancouver), the suggestion of a possible link between these findings is compelling. This analysis provides a foundation to hypothesize a relationship between *differential industrial structures* and *FGF tracking results*. The conclusion that follows summarizes the key points emerging from the above analysis and links these results to salient literature and possible interpretations.

Summary and conclusion

We find that the track records of Canadian FGFs demonstrate the ongoing importance of these firms in a regional development context. These tracking results show that FGFs remain in their original, host metropolitan area in numbers far surpassing other possible firm outcomes, which include relocation, acquisition, and firm discontinuation. A complementary analysis of these tracking results by metropolitan area indicates that such FGF continuity has a substantial geographic dimension, with certain Canadian metropolitan areas (including Toronto and Vancouver) retaining many of their FGFs locally, while other metropolitan areas (including Montréal and Calgary) experience substantially higher rates

Table 8. NAICS sectors for large firms in Calgary and Vancouver, 2005.*

Metropolitan area	NAICS sector	Number of large firms	Sector percentage of all metro large firms
Calgary			
	2111: Oil and Gas Extraction	45	50.6
	2131: Support Activities for Mining	10	11.2
	2211: Electrical Power Generation	6	6.7
	2362: Industrial Construction	3	3.4
	3251: Petrochemical Manufacturing	3	3.4
	5311: Real Estate	3	3.4
	2212: Natural Gas Distribution	2	2.2
	4451: Food Retailers	2	2.2
	5171: Wired Telecommunications	2	2.2
	17 Additional Sectors	1 each	1.1 each
	Calgary metropolitan area total	***89***	***100.0***
Vancouver			
	2122: Metal Ore Mining	7	8.6%
	3221: Pulp and Paper	6	7.4%
	4236: Electronics Wholesalers	5	6.2%
	5221: Banking and Savings	5	6.2%
	3342: Telecommunications Equipment	4	4.9%
	5231: Investments and Securities	4	4.9%
	2212: Natural Gas Distribution	3	3.7%
	4541: Non-Store Retailing	3	3.7%
	5311: Real Estate	3	3.7%
	7221: Restaurants	3	3.7%
	2362: Industrial Construction	2	2.5%
	3254: Bio-Pharmaceuticals	2	2.5%
	3341: Electronics Manufacturing	2	2.5%
	4233: Building Product Wholesalers	2	2.5%
	4244: Grocery Wholesalers	2	2.5%
	4413: Automotive Parts Stores	2	2.5%
	5511: Diversified Companies	2	2.5%
	17 additional sectors	1 each	1.2% each
	Vancouver metropolitan area total	***81***	***100.0%***

Source: Financial Post (2006).

Note: *Table data refer to corporate status as of 2005, the final year of the study's FGF tracking analysis.

of business failure and relocation. An in-depth analysis of FGFs by sector in Calgary and Vancouver indicates that Vancouver's successful FGF community is clustered in high-technology and food sectors that were not closely related, while Calgary's less successful FGFs are fall almost exclusively into one very large energy cluster or are on their own in NAICS industry categories with no local counterparts.

These findings suggest a number of observations, hypotheses, and conclusions. First, the results demonstrate that FGFs have promise as development catalysts in the Canadian context due to their overall longevity and locational stability. In the years following their most rapid growth, many Canadian FGFs continue in independent operation for extended periods in their original host cities. This finding reinforces the importance of continued attention to FGFs from a regional development perspective and supports the rationale for FGF interest outlined earlier in this article, linking FGFs with HQ city system evolution (Rice & Lyons, 2010). A fruitful linkage in this regard would place FGF development in a context that also includes the development of entire industries, such as those examined within Markusen's (1985) seminal "Profit Cycle" theory, Audretsch's (1995) research on innovation and industry evolution, and more broadly within the growing literature related to techno-economic paradigms (Cantwell & Santangelo, 2000; Geels, 2006; Hayter, 2008). The 'stickiness' of FGF activity demonstrated here suggests that FGFs are a key contributor to HQ cluster development.

A second observation, focused on the western Canadian context, is that attention to regional industrial composition may provide insight into FGF development. This observation relates FGF development to the seminal regional industrial structure comparisons of Chinitz (1961) and to more recent contributions related to regional industrial "lock-in" and flexibility (Christopherson & Clark, 2007; Grabher, 1993; Treado & Giarratani, 2008). The Detroit example cited earlier is relevant to this connection, as it indicates that specialization can propel growth through a region's corporate emergence, but if unaccompanied by diversification, such industrial focus can become a weakness as the region matures. Vancouver appears to be avoiding this trap at present. Our results suggest that Vancouver's metropolitan business community has been markedly more successful than Calgary's in FGF development because of the differential industrial structures characterizing the two regions. The high degree of distinction between the two regions with regard to industrial concentration and development of focused groups of FGFs in related industries is a particular point of contrast. These structural indicators are especially relevant in two dimensions: first, in terms of regional opportunities for collaboration among related but noncompeting firms, and second regarding competition with firms providing directly overlapping products and services inside well-defined industries. This evidence suggests that it is important for regional business communities both to build a critical mass of related businesses (as could be hypothesized for Vancouver) and to avoid what might be conceived as too high a degree of firm overlap inside any single industry (as observed in Calgary).

A third conclusion is that Vancouver and Calgary's large-firm industrial structures have parallels with each region's FGF structures. Our results imply that overall metropolitan industrial structure, and not just FGF industrial composition, may play an important role in shaping FGF development results. Our results here indicate that resource-related sectors are important in both cities, but that Vancouver's business community appears to have diversified beyond resources and is growing in sectors not directly related to the city's historical resource base. Vancouver's overall corporate expansion, while not as rapid as Calgary's, can hypothesized as being potentially more sustainable, based on the Vancouver business community's diverse FGF industrial distribution and regional possibilities for interfirm complementarities. At the same time, it would appear that the key challenge for Calgary's FGF community parallels the primary challenge for the city's largest firms: can the city parlay its massive resource-based growth into complementary (and even unrelated) development in other industries? The evidence here indicates that

Calgary's business community has yet to succeed in this dimension, but it is an open question as to whether this may occur in the future.

A fourth and final observation focuses on the other pair of major corporate centers, Toronto and Montréal. Our findings provide some insight into the mechanisms responsible for Toronto and Montréal's contrasting business trajectories over the past 30 years. Since 1970, when Toronto and Montréal were near equals in corporate status at the top of the Canadian urban hierarchy, Toronto has emerged as the undisputed national business and financial center for Canada, while Montréal has devolved into the country's second-ranking corporate center (Beckstead & Brown, 2006; Rice & Pooler, 2009; Semple, 1996). Prominent explanations in the academic and popular literature for this corporate shift focus on the emergence of the Quebec independence movement and associated political turmoil (Black, 2012; Rice & Semple, 1993; Semple, 1996). However, our findings on FGFs in these two cities add depth to our understanding of the situation. The evidence offered here demonstrates that positive economic development conditions have continued to exist in Montréal, as a sizable number of FGFs emerged in the region through the 1987–2005 study period (ranking the city third in this regard within Canada, ahead of Calgary). The negative component of our findings that the majority of these firms did not continue their operations in Montréal may suggest that Montréal's business community has encountered a very specific challenge: FGFs generated in the region have had great difficulty continuing to develop while located in Montréal.[4] Toronto, by comparison, has continued to see its many FGFs grow and succeed locally, paralleling the continued success of its large-firm community. Although more research is necessary to understand the issues raised here, our findings provide a focus for businesses and public policymakers interested in addressing the economic imbalance that has emerged between Montréal and Toronto since the early 2000s.

As mentioned earlier, more investigation is necessary to expand the findings presented here. Earlier discussion acknowledged that our study notes parallel trends in FGF development and industrial structure and argues that a linkage between the two may exist. More evidence is needed to explore the potential existence and nature of such a linkage. More in-depth analysis of the FGFs and large firms involved, including case studies of specific metropolitan industries (such as Calgary's energy complex and Vancouver's food and high-technology clusters), would be a logical next step. Continued monitoring of FGF development nationwide and by metropolitan area is also necessary, as more and better data (including longer-term tracking, from business birth to potential firm emergence as national players) would help to better define the paths by which regional HQ complexes emerge and mature. Finally, survey-based research is a needed complement to the database-driven analysis in the present study. Such extensions are important, as research in this field is central to explaining the location and distribution of HQ development and providing public- and private-sector decision-makers with tools to assess, attract, and retain elite corporate activities and their associated benefits.

Acknowledgements

We acknowledge the constructive comments of the anonymous reviewers. Any remaining errors are ours alone.

Funding

This work was completed with Canadian Studies Research Grant assistance from the Government of Canada/avec l'appui du gouvernement du Canada.

Notes

1. FGFs have been studied under a variety of names, including "gazelle" (Birch, 1981; Stam, 2005), "next wave firm" (Rice, 2004; Rice & Lyons, 2007), "10 percenter" (Storey, 1994), and "high-growth firm" (Moreno & Casillas, 2007). These terms have a wide range of definitions that vary in focus and in degree of specificity, but all center broadly on rapid business growth. Rather than focusing on a highly specific definition, our study uses FGF as an overall designation to denote the fastest-growing businesses in a given economy.
2. It is important to acknowledge that firm relocation (or acquisition together with relocation) may be in the best interests of some businesses, such as a firm whose growth is constrained by a small local market or labor force but which has the potential to expand in a larger urban area. Regardless, given the regional development focus of this article, we interpret high levels of firm continuity as having some positive connotation for the region and its business community.
3. A likely factor in Vancouver's success may relate to its access to markets beyond North America, particularly in the Asia-Pacific region.
4. It is possible to interpret firm discontinuation as a sign of regional strength. Business failure in Montréal could be seen as the result of a dynamic and competitive environment in which only strong firms are surviving, and not necessarily the result of an inherent problem with the regional economy. However, to convincingly make such an argument, more evidence would be needed to account for the much greater success of FGFs in Toronto's highly competitive business community and to provide an explanation for the degree to which Montréal-based FGFs fall in the "defunct/no records" category at a rate far above the national average (38.1% of FGFs in Montréal vs 23.7% nationally).

References

Acs, Zoltan J., Parsons, William, & Tracy, Spencer (2008). *High impact firms: Gazelles revisited.* Washington, DC: Small Business Administration.

Almus, Matthias. (2002). What characterizes a fast growing firm? *Applied Economics, 34*(12), 1497–1508.

Audretsch, David B. (1995). *Innovation and industry evolution.* Cambridge, MA: The MIT Press.

Baldwin, John R., Bian, Lin, Dupuy, Richard, & Gellatly, Guy. (2000). *Failure rates for new Canadian firms: New perspectives on entry and exit.* (Catalogue No. 61-526-XIE). Ottawa, ON: Statistics Canada.

Beckstead, Desmond, & Brown, W. Mark. (2006). *Head office employment in Canada, 1999 to 2005* (Canadian Economic Observer, July. Catalogue No. 11-010-XIB). Ottawa, ON: Statistics Canada.

Beyers, William B. (2002). Services and the new economy: Elements of a research agenda. *Journal of Economic Geography, 2*(1), 1–29.

Birch, David L. (1981). Who creates jobs? *The Public Interest, 65*, 3–14.

Birch, David L. (1987). *Job creation in America: How our smallest companies put the most people to work.* New York, NY: The Free Press.

Bishop, K., Mason, G., & Robinson, C. (2009). *Firm growth and its effects on economic and social outcomes: Literature and statistical review.* Report to the National Endowment for Science Technology and the Arts. London: NIESR.

Black, Conrad. (2012, October 6). As Quebec decays, Toronto seizes greatness. *National Post*, Section A18.

Britton, John N. H., & Gilmour, James M. (1978). *The weakest link: A technological perspective on Canadian industrial underdevelopment.* Ottawa, ON: Science Council of Canada.

Canadian Newspaper Services International. (1990–2006). *The blue book of Canadian business.* Toronto, ON: Canadian Newspaper Services International.

Cantwell, John, & Santangelo, Grazia D. (2000). Capitalism, profits and innovation in the new techno-economic paradigm. *Journal of Evolutionary Economics, 10*, 131–157.

Card, David, Hallock, Kevin F., & Moretti, Enrico (2010). The geography of giving: The effect of corporate headquarters on local charities. *Journal of Public Economics, 94*(3), 222–234.

Chamberlin, Tyler, & de la Mothe, John (2004). Northern light: Ottawa's technology cluster. In David A. Wolfe (Ed.), *Clusters old and new* (pp. 213–234). Montreal, QC: McGill-Queen's University Press.

Chinitz, Benjamin (1961). Contrasts in agglomeration: New York and Pittsburgh. *The American Economic Review, 51*(2), 279–289.
Christopherson, Susan, & Clark, Jennifer (2007). *Remaking regional economies: Power, labor, and firm strategies in the knowledge economy.* London: Routledge.
Coad, Alex. (2009). *The growth of firms: A survey of theories and empirical evidence*. Cheltenham: Edward Elgar.
Delmar, Frédéric, Davidsson, Per, & Gartner, William B. (2003). Arriving at the high growth firm. *Journal of Business Venturing, 18*(2), 189–216.
Dun and Bradstreet. (1987–2007). *Canadian key business directory.* Toronto, ON: Dun and Bradstreet Canada.
Dun and Bradstreet. (1990–2006). *Canadian key business directory.* Toronto, ON: Dun and Bradstreet Canada.
Financial Post. (1987). *Financial post 500*. Toronto, ON: The Financial Post Company.
Financial Post. (1990–2006a). *FP corporate surveys*. Toronto, ON: Postmedia Network.
Financial Post. (1990–2006b). *FP investor reports*. Toronto, ON: Postmedia Network.
Financial Post. (1990–2006c). *FP predecessor & defunct*. Toronto, ON: Postmedia Network.
Financial Post. (2006). *Financial post 500*. Toronto, ON: Postmedia Network.
Financial Post. (2007). *Financial post 500*. Toronto, ON: Postmedia Network.
Geels, Frank W. (2006). Co-evolutionary and multi-level dynamics in transitions: The transformation of aviation systems and the shift from propeller to turbojet. *Technovation, 26*(9), 999–1016.
Gong, Hongmian, & Wheeler, James O. (2007). Geographic changes in the metropolitan corporate South, 1995-2003. *Southeastern Geographer, 47*(1), 155–169.
Goodwin, William. (1965). The management center in the United States. *Geographical Review, 55*(1), 1–16.
Grabher, Gernot. (1993). The weakness of strong ties: The lock-in of regional development in the Ruhr area. In Gernot Graber (Ed.), *The embedded firm: On the socio-economics of industrial networks* (pp. 255–277). London: Routledge.
Graves, William, & Kozar, J. (2010). Blending Southern culture and international finance: The construction of a global money center. In William Graves & Heather A. Smith (Eds.), *Charlotte, NC: The global evolution of a New South City* (pp. 87–101). Athens: University of Georgia Press.
Green, Milford B. (1993). A geography of institutional stock ownership in the United States. *Annals of the Association of American Geographers, 83*(1), 66–89.
Green, Milford B. (2004). Venture capital investment in the United States 1995–2002. *The Industrial Geographer, 2*(1), 2–30.
Green, Milford B., & Semple, R. Keith (1981). The corporate interlocking directorate as an urban spatial information network. *Urban Geography, 2*(2), 148–160.
Hayter, Roger (2008). Environmental economic geography. *Geography Compass, 2*(3), 831–850.
Henrekson, Magnus, & Johansson, Dan (2010). Gazelles as job creators: A survey and interpretation of the evidence. *Small Business Economics, 35*(2), 227–244.
Holloway, Steven R., & Wheeler, James O. (1991). Corporate headquarters relocation and changes in metropolitan dominance, 1980-1987. *Economic Geography, 67*, 54–74.
Honjo, Yuji (2004). Growth of new start-up firms: Evidence from the Japanese manufacturing industry. *Applied Economics Letters, 11*(1), 21–32.
Jean, S. (2013, June 2). Mining their business: Texas, other states rekindle battle for one another's companies, jobs. *The Dallas Morning News*, Sections 1D and 6D.
Knoben, Joris (2011). The geographic distance of relocation search: An extended resource-based perspective. *Economic Geography, 87*(4), 371–392.
Lee, M. (2010, June 16). Council approves tax rebates. *Fort Worth Star-Telegram*, Sections 1C–2C.
Lord, J. Dennis (1992). Geographic deregulation of the U.S. banking industry and spatial transfers of corporate control. *Urban Geography, 13*(1), 25–48.
Markman, Gideon D., & Gartner, William B. (2002). Is extraordinary growth profitable? A study of Inc. 500 high-growth companies. *Entrepreneurship Theory and Practice, 27*(1), 65–75.
Markusen, Ann R. (1985). *Profit cycles, oligopoly, and regional development*. Cambridge, MA: The MIT Press.
Meyer, Stephen, & Green, Milford B. (2003). Headquarters in Canada: An analysis of spatial patterns. *Urban Geography, 24*(3), 232–252.

Mitchelson, Ronald L., & Wheeler, James O. (1994). The flow of information in a global economy: The role of the American urban system in 1990. *Annals of the Association of American Geographers*, *84*(1), 87–107.
Moreno, Ana M., & Casillas, José C. (2007). High-growth SMEs versus non-high-growth SMEs: A discriminant analysis. *Entrepreneurship & Regional Development*, *19*(1), 69–88.
O'Hagan, S., Rice, Murray D., & Green, Milford B. (2008). North American corporate directors and educational affiliations: A geographical analysis. *Geography Research Forum*, *28*, 59–81.
Parker, Simon C., Storey, David J., & van Witteloostuijn, Arjen (2010). What happens to gazelles? The importance of dynamic management strategy. *Small Business Economics*, *35*(2), 203–226.
Poremba, David Lee (2003). *Detroit: A motor city history*. Mount Pleasant, SC: Arcadia.
Porter, Michael E. (2000). Locations, clusters and company strategy. In Gordon L. Clark, Maryann P. Feldman, & Meric S. Gertler (Eds.), *The Oxford handbook of economic geography* (pp. 253–274). Toronto, ON: Oxford University Press.
Profit Magazine. (1988–2003). *Profit 500*. Toronto, ON: Rogers Media.
Quante, Wolfgang (1976). *The exodus of corporate headquarters from New York City*. New York, NY: Praeger.
Rice, Murray D. (2004). Regional and sectoral growth in Canada's emerging economy. *Canadian Journal of Regional Science*, *27*(2), 237–254.
Rice, Murray D. (2006). Dominant national centers: A comparative analysis of the headquarters communities of New York and Toronto. *Industrial Geographer*, *4*(1), 28–50.
Rice, Murray D., & Lyons, Donald I. (2007). Consistent business generators: Urban leaders in Canada's next wave. *Canadian Journal of Regional Science*, *30*(1), 91–112.
Rice, Murray D., & Lyons, Donald I. (2010). Geographies of decision-making and control: Development, applications, and future directions in headquarters location research. *Geography Compass*, *4*(4), 320–334.
Rice, Murray D., & Pooler, James A. (2009). Subsidiary headquarters: The urban geography of the 'second tier' of corporate decision-making activity in North America. *Urban Geography*, *30*(3), 289–311.
Rice, Murray D., & Semple, R. Keith (1993). Spatial interlocking directorates in the Canadian urban system, 1971-1989. *Urban Geography*, *14*(4), 375–396.
Rice, Murray D., Tierney, Sean, O'Hagan, Sean, Lyons, Donald, & Green, Milford B. (2012). Knowledge, influence, and firm-level change: A geographic analysis of board membership associated with Canada's growing and declining businesses. *Geoforum*, *43*(5), 959–968.
Semple, R. Keith (1985). Toward a quaternary place theory. *Urban Geography*, *6*(4), 285–296.
Semple, R. Keith (1996). Quaternary places in Canada. In John N. H. Britton (Ed.), *Canada and the global economy: The geography of structural and technological change* (pp. 352–373). Montréal, QC: McGill-Queen's University Press.
Stam, Erik (2005). The geography of gazelles in the Netherlands. *Tijdschrift voor Economische en Sociale Geografie*, *96*(1), 121–127.
Stam, Erik (2010). Growth beyond Gibrat: Firm growth processes and strategies. *Small Business Economics*, *35*(2), 129–135.
Steed, Guy P. F. (1982). *Threshold firms: Backing Canada's winners*. Ottawa, ON: Science Council of Canada.
Stephens, John D., & Holly, Brian P. (1981). City system behavior and corporate influence: The headquarters location of US industrial firms, 1955–75. *Urban Studies*, *18*(3), 285–300.
Storey, David (1994). *Understanding the small business sector*. London: Routledge.
Treado, Carey D., & Giarratani, Frank (2008). Intermediate steel-industry suppliers in the Pittsburgh region: A cluster-based analysis of regional industrial resilience. *Economic Development Quarterly*, *22*(1), 63–75.
Walcott, Susan M., & Wheeler, James O. (2001). Atlanta in the telecommunications age: The fiber-optic information network. *Urban Geography*, *22*(4), 316–339.
Wheeler, James O. (1988a). The corporate role of large metropolitan areas in the United States. *Growth and Change*, *19*(2), 75–86.
Wheeler, James O. (1988b). Spatial ownership links of major corporations: The Dallas and Pittsburgh examples. *Economic Geography*, *64*(1), 1–16.
Wheeler, James O. (1990). The new corporate landscape: America's fastest growing private companies. *Professional Geographer*, *42*(4), 433–444.
Woodford, Arthur M. (2001). *This is Detroit: 1701–2001*. Detroit, MI: Wayne State University Press.

Interaction of corporate and urban systems: accumulation of intangible assets

Sam Ock Park[a,b]

[a]*Department of Geography, Seoul National University, Seoul, Korea;* [b]*Department of Public Administration, Gachon University, Seongnam City, Korea*

This study develops a conceptual framework of the relationship between corporate systems and urban systems in the process of intangible asset accumulation and conducts case studies involving two distinctive and contrasting global firms. Intangible assets are important in maintaining the competitive advantage of cities and regions. The accumulation of local intangibles is the result of the evolutionary development of the relationship between the corporate system and the urban system. The results of the study suggest that the intangible assets of firms are critical in ensuring their competitiveness within a knowledge-based information society.

Introduction

Globalization has been one of the significant megatrends reshaping the world's economic spaces during the last three decades. The increase in foreign direct investments (FDIs), which exceeds the pace of growth in international trade, has dramatically shaped and reshaped the global economy, particularly the emerging East Asian economies. The progress of a knowledge-based economy and an information society has overlapped with globalization and resulted in the remarkable global flows of intangible assets, which include knowledge, human capital, technological know-how, and brand identities—as well as of tangible assets through FDIs and international trade of products and resources. The progress of a knowledge-based information society is coupled with two other phenomena: (1) the relative decrease in the importance of physical resource endowment as drivers of regional development and firm growth and (2) the increase in the importance of intangible assets related to personal knowledge for both firm and region (Surinach & Moreno, 2012).

The intangible asset accumulation of firms and regions has become a critical factor for attaining competitiveness. Specifically, the accumulation of a kind of intangible asset in a given city or region is important in attracting FDIs for industrializing countries because the externalities produced by knowledge spillover from firms can influence the accumulation of regional intangible assets. Subsequently, accumulation results in a positive relationship between firm and region in the knowledge-based economy (Artis, Miguelez, & Moreno, 2012). Accordingly, the creation, accumulation, and flow of intangible assets in the global space economy have become significant drivers of change in global economic spaces along with the competitiveness and productivity of firms and regions. In the processes that constitute the intangible asset accumulation of a firm, an organization's surrounding city and regional context becomes important in the acquisition and integration of external knowledge beyond the internal knowledge base of the firm. These processes reflect the interaction of the firm with diverse regional settings, reproducing variations in region-specific knowledge and confirming the importance of the regional scale as a fundamental site of innovation (Asheim, Boschma, & Cooke, 2011).

Traditionally, the interaction between firm and region for industrial location has been integrated in economic and urban geography theories. The recent history of firm–region interaction has revealed a shift in emphasis from tangible assets to firms' access to intangible assets (*Regional Studies*, 2012, Vol. 46.10). In corporate geographic research, the focus has shifted from manufacturing activities to knowledge and information flows and corporate networks established with financial institutions (Gong & Wheeler, 2002; Park & Wheeler, 1983; Wheeler, 1981, 1986, 1999). James O. Wheeler (1986, 1988a, 1988b) emphasized the corporate role of the relationship between firms and urban areas in the US urban systems.

In honor of James O. Wheeler, the present study extends the relationship between corporate systems and urban systems to the global space economy beyond the United States, as well as to intangible assets, by considering the two global megatrends—globalization and the rise of a knowledge-based information society. The relationship between corporate hierarchies and urban hierarchies is examined, in the context of intangible asset accumulation, as a process aimed at increasing the competitiveness of firms and regions. Several studies have explored the importance of intangible assets in the competitiveness and productivity of firms (e.g., Artis et al., 2012; Bianchi & Labory, 2004; Marrocu, Paci, & Pontis, 2012; O'Mahony & Vecchi, 2009; Surinach & Moreno, 2012). The role of intangible assets in regional economic growth has been broadly examined in a recent special issue of *Regional Studies* (2012). However, the role of corporate and urban systems in intangible asset accumulation that is aimed at increasing the competitiveness of both firm and region has not been thoroughly investigated. Based on Wheeler's pioneering research on corporate systems and urban hierarchy in terms of headquarter location in the United States, this study is regarded as an extension of his work particularly because of its focus on intangible assets.

The major purpose of this study is to investigate the conceptual framework of the processes of intangible asset accumulation through the interaction of corporate and urban systems. To understand the interaction between corporate and urban systems as well as intangible asset accumulation, this study performs a historical analysis of two distinctive and contrasting Korean firms comprising more than 200 establishments. The analysis incorporates various statistics and reports documenting the firms' spatial and structural strategies for growth. In-depth interviews are also conducted to clarify such interactions and the processes of intangible asset accumulation. Qualitative methods play an important role in understanding the complex plurality of contexts and the existing networks between firms and their environment (Birkinshaw, Brannen, & Tung, 2011). Section 2 examines the firm–region relationship with regard to intangible assets, whereas Section 3 discusses the conceptual framework of the interaction between corporate and urban systems. Section 4 analyzes the two cases, Samsung Electronics Corporation (SEC) and Lutronic Corporation, based on the conceptual framework. The final section presents the concluding remarks.

The firm–region relationship and intangible assets

The firm–region relationship has become a major focus of geography research since the publication of the works of traditional location theorists such as Alfred Weber (1929) and August Lösch (1952). However, most of the early studies on the firm–region relationship have concentrated on the organization of tangible assets in location factors, such as access to material sources and markets, transportation, and labor. Firms interact with regions in the process of making location decisions that consider accessibility to input materials and markets as well as in securing cheap labor. In firm–region interaction, transportation cost is a critical factor in the spatial organization of production activities. Along with the progress of the knowledge-based economy during the last few decades, the importance of intangible assets such as knowledge, technological innovation, and brand recognition in firm–region interaction has increased, as these assets can improve productivity and generate competitiveness (Bianchi & Labory, 2004; Fey & Birkinshaw, 2005; Surinach & Moreno, 2012).

An intangible asset is defined by its nonphysical nature. It is regarded as a "non-material factor to enterprise performance in the production of goods or the provision of services" that may be "expected to generate future economic benefits to the entire firm or individual that control the deployment" (Tommaso, Paci, & Schweitzer, 2004, p. 76). Intangible assets are generally considered as "all the possible sources of competitive advantages for firms that have no physical substance" (Tommaso et al., 2004, p. 77). In the financial report of a firm, identifiability is emphasized to distinguish an intangible asset from the broader conceptualization of "goodwill" as an asset. Intangible assets are widely recognized as the driving force of the productivity growth of the knowledge-based economy and are becoming increasingly crucial for firm survival and productivity (Fey & Birkinshaw, 2005; Marrocu et al., 2012). Furthermore, intangible assets are regarded as sources of competitive advantage not just for firms but also for cities and regions in the globalized economy (Bianchi & Labory, 2004; Kramer, Revilla Diez, Marnelli, & Iammarino, 2009). Intangible assets can be categorized into different bundles. Anson (2007) categorized intangibles into separate bundles for marketing, IT, technical capacity, and so on. For convenience, these bundles are used conceptually in the present work to describe the relationship of firms with urban systems.

Understanding firm behavior when seeking and exploiting intangible assets is helpful in studying the firm–region relationship in the knowledge-based economy. Traditionally, the FDIs are regarded as the exploitation or transfer of firm-specific advantages, assuming that the firms possess certain types of proprietary resources that can be exploited in the host country (Hymer, 1976; Makino, Lau, & Yeh, 2002). However, recent international business studies have suggested that firms invest in foreign countries to develop their firm-specific advantages or to seek intangible assets that are required to make use of their firm-specific assets in a host country (Dunning, 1995; Kumar, 1998). Firms engage in FDIs to accumulate intangible assets by learning and gaining access to the unique assets available in a host country. Another study has suggested that firms of newly industrialized economies gain access to established brand names, novel product technologies, and extensive distribution networks by investing in advanced countries (Kumar, 1998). Such asset-seeking and asset-exploiting behaviors are consistent with one of the long-standing ideas in organization science, that is, the long-term success of an organization depends on its ability to explore new competencies and exploit its current capabilities (Raisch, Birkinshaw, Probst, & Tushman, 2009). Raisch et al. (2009, p. 685) have suggested that superior performance is expected from ambidextrous organizations "that are capable of simultaneously exploiting existing competencies and exploring new opportunities." In this study, we assume that firms interact with various cities through asset-seeking and asset-exploiting behaviors with regard to intangible assets by utilizing the characteristics of cities and regions in the global space economy.

In accordance with their asset-seeking behavior, firms carefully select their local area of investment by considering the local industrial clustering, regional innovation systems, local culture, and the environment based on their investment functions. For example, the presence of supplier networks specializing in input materials, labor, and so on in the local area will be a highly important consideration in site selection for manufacturing plants, whereas the existence of innovation networks and pools of scientists and engineers will be the crucial factor for determining the location of R&D institutes. The characteristics of the local labor market, regional technology, and knowledge bases will be carefully considered in investments in R&D center—and different regions will have different knowledge and technology bases. Therefore, exploring or seeking intangible assets is important in investments in R&D centers. In production investment, the asset-exploiting behavior may be considered in establishing local production networks. Firms exploit knowledge through foreign sales and marketing, on the one hand, and seek knowledge to develop new products and processes, on the other (Dunning, 1995; Makino et al., 2002). However, the asset-seeking and asset-exploiting behaviors of firms cannot be regarded as independent of each other. Firms move from asset exploiting to asset seeking and vice versa in their investments. Through the continuous cycle of asset exploiting and asset seeking, firms can accumulate intangible assets. In the knowledge-based information society, multinational firms (MNFs) continuously progress through investments of different functions in various cities and regions. Accordingly, MNFs are hypothesized to be capable of concurrently pursuing both asset-seeking and asset-exploiting behaviors by differentiating the cities and regions in the global space economy. Schematically, the interaction and networks of firms with suppliers, customers, competitors, universities, governments, and communities result in the accumulation of knowledge, technology, human capital, brand value, and so on. Knowledge spillover through the networks within firms and inter-organizational networks is an important component of the intangible asset accumulation of firms.

Cities and regions are fields of the collective action of individuals, organizations, and social institutions. Actors in cities and regions mobilize heterogeneous resources from both private and public institutions. With these collective actions and mobilization, cities and regions function as the setting in which physical, human, and social capital can be combined to create competitive environments where high productivity, innovativeness, and adaptability coexist. Cities and regions evolve through the accumulation of local intangible assets. Creating local knowledge spillover channels, accessing knowledge capital, identifying local knowledge bases for intangible asset seeking, and determining the local market base for intangible asset exploitation are the processes involved in the accumulation of regional intangible assets. In these accumulation processes, the socio-economic characteristics of firms, such as embeddedness, networks, and reciprocity, are important (Dicken & Malmberg, 2001; Taylor & Asheim, 2001).

Firm–region interaction can develop through the accumulation of regional assets. Similarly, intangible assets are continuously accumulated in certain regions where an industrial cluster, regional innovation system, or qualified local labor market is well developed (Kramer, Marinelli, Iammarino, & Revilla Diez, 2011). Firms can concentrate in environments that are favorable for seeking new knowledge and information on new products and process technology, pooling qualified labor, and establishing innovation networks. Industrial clusters are the result of close formal and informal networks that aim to acquire information and knowledge of new products and process innovations, marketing, and sales beyond mere material linkages. Beyond the material linkages of supplier and customer relations, intangible asset accumulation in industrial clusters is important in locating firms within the clusters. The establishment of innovation systems in a region through continuous engagement with formal and informal networks among firms, universities, and R&D centers, as well as the existence of favorable institutions and culture and social capital, can attract firms that seek knowledge and technology assets for innovation and R&D activities. Industrial clusters and innovation centers can accumulate regional intangible assets through the continuous interaction of firms, whereas firms can accumulate intangible assets through their participation in formal and informal networks with regions.

Interactions between corporate systems and urban systems

The systems approach is helpful in clearly defining the firm–region relationship. Dicken and Malmberg (2001) have suggested the firm-territory nexus as a schematic framework in which industrial and governance systems are interlinked. Production and innovation systems are considered in industrial systems, and firms are regarded as networks of corporate organizational processes interacting with regional industrial systems. This schematic framework is useful in understanding the complex firm–region relationship theoretically. The present study extends the previous work of Wheeler by defining the firm–region relationship through the interdependencies of corporate systems and urban systems (Wheeler, 1986, 1988a, 1988b).

Exploring the relationship between corporate and urban systems is useful in identifying firm–region relationships with regard to intangible assets because different firm units have different functions and are linked with different types of regions or cities. Corporate systems include networks among the separate functional units of the firm and networks with other external economic actors, such as suppliers, customers, competitors/cooperators, communities, universities, and the government (Figure 1). Networks include the flows of both tangibles and intangibles. Based on this theoretical framework, three hypotheses on the interaction of corporate systems with urban systems merit investigation.

First, we hypothesize that the hierarchy of a corporate system is related to the hierarchy of an urban system in the flow of intangible assets within the corporate system (H1). A firm interacts with different functional units within the organization for both tangibles and intangibles; this relationship can be regarded as an intra-firm network. In this regard, a corporate system can be defined as the interaction among diverse functional units of a firm in the economic space. Specifically, the functional units of a firm, such as its headquarters, regional headquarters, production plants, R&D centers, and sales offices, are parts of a corporate system that interact within the firm for the sharing, accumulation, and spillover of knowledge, as well as for networking and for the flow of tangibles. The flow of tangibles and intangibles such as knowledge, technology, and brand in joint ventures, subsidiaries, affiliates, and distribution channels can also be classified as intra-firm networks. An MNF tends to locate its headquarters and regional headquarters in the capital region or in the largest metropolitan area because these areas are the hubs of intangible assets within the corporate system. As the gathering and flow of knowledge and information are important for R&D centers, these centers also tend to locate in the capital region or in the knowledge centers of a nation. However, the network of intangible assets may not be as important as that of tangible assets for production units; thus,

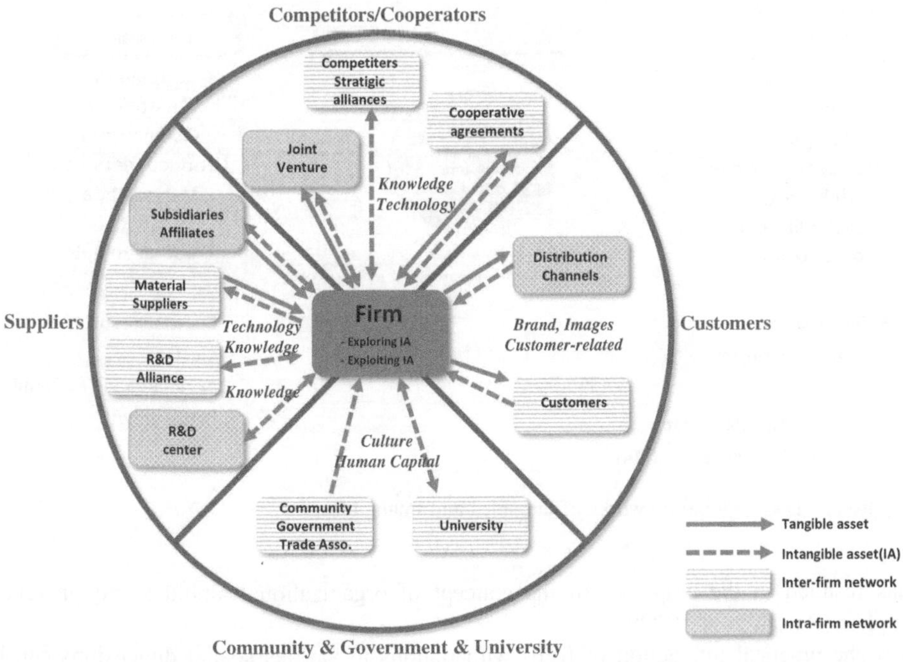

Figure 1. Networks of firm.

manufacturing plants can be located even in peripheral cities. In gathering intangible assets in the inter-organizational network, firms take into consideration the different external environments of cities based on their varying functions. A firm interacts with the external environment through networks with customers, suppliers, competitors, universities, and other institutions located in diverse regions; this relationship can be regarded as an inter-organizational network (Figure 1). However, network characteristics may be different depending on the type of the firm's operating units.

Second, we hypothesize that in seeking and exploiting intangible assets, agglomeration and clusters are important at the local level, whereas transnational cluster networks and temporary clusters are important at the global level (H2). Inter-organizational networks are established at various spatial scales for the purpose of seeking and exploiting intangible assets (Figure 2). Close interactions with suppliers, customers, competitors, producer services, universities, and the government within a local area are important in seeking and exploiting intangible assets and in lowering transaction costs. These interactions can be developed at international scales, with various types of global networks comprising diverse economic actors (Figure 2).

Third, we hypothesize that most MNFs in a knowledge-based economy tend to explore and exploit intangible assets simultaneously (H3). We previously emphasized the exploitative behavior of MNFs in developing countries and their exploratory behavior in advanced countries. However, most MNFs have corporate units in both advanced and developing economies, and both behaviors can be simultaneously present. The exploration and exploitation of intangible assets are continuous and complex cycles that can be observed simultaneously at a given time in the global economic space. This complexity

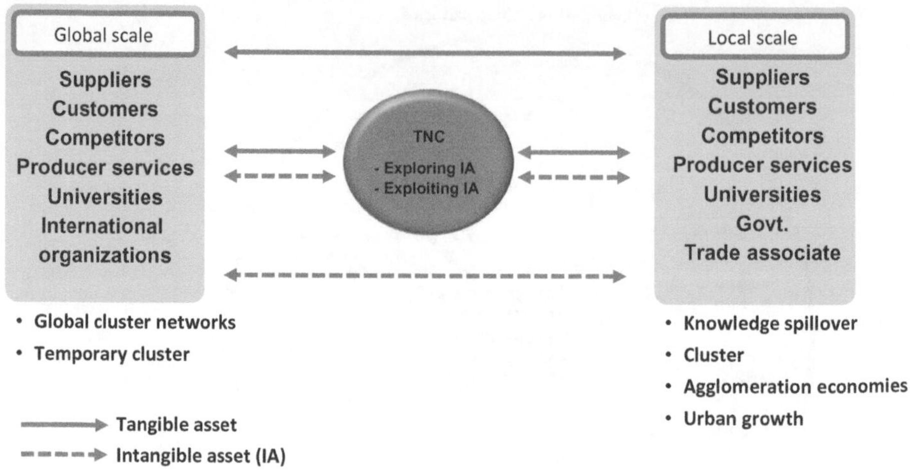

Figure 2. Local–global networks of tangibles and intangibles.

has resulted in the emphasis of the concept of organizational ambidexterity in recent studies (Raisch et al., 2009).

The practical interaction of firms with regions in various spatial dimensions can be described in terms of urban systems because the individual units of a firm are actually located in diverse cities within the regional, national, and global economic spaces. The headquarters of a firm control and interact with all the firm's functional units across different locations. The headquarters also interacts with its external environment. On the one hand, firms tend to locate their headquarters in high-ranking metropolitan areas in the national urban system to maintain effective control and efficient interaction with other actors within the city in relation to intangible assets. Examples of such areas include Seoul in Korea, Tokyo in Japan, and New York in the United States. On the other hand, firms tend to locate their branch plants in peripheral cities or in developing countries where production networks for both tangibles and intangibles can be established efficiently. Similarly, different units of corporate systems require different locational considerations. These units tend to locate in and interact with different cities within the national urban systems of global economic spaces to accumulate and increase intangible assets.

The individual units of corporate systems interact with cities in the urban systems to accumulate intangible assets—such as knowledge and technical bundles, networking bundles, marketing bundles, and cultural bundles. In corporate systems, the relationship among the different functional units of firms located in different regions and occupying different positions within the organizational structure is embedded within the spatiality of such firms (Dicken & Malmberg, 2001). Cities or regions accumulate intangible assets through the interaction among actors in the local area to share knowledge, establish global networks, develop local social capital and adherence to local culture, and establish local innovation systems. Intangible asset accumulation in cities or regions can have a positive effect on firms that are exploring and seeking to develop intangible assets in the local area. Specifically, the exploration of intangible assets results in the accumulation of such assets of firms and vice versa, as shown in Figure 3, which likewise presents the relationship between corporate systems and urban systems.

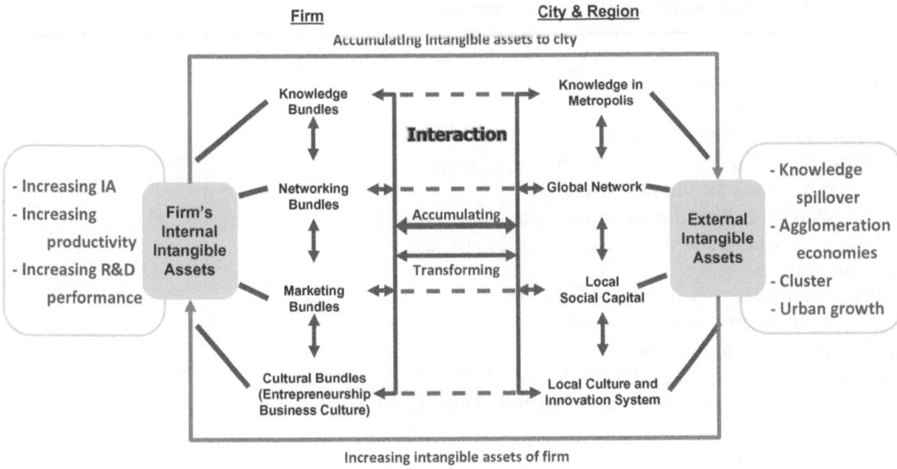

Figure 3. Accumulating intangible assets through interactions between firm and city.

Case studies: Samsung Electronics Corporation and Lutronic Corporation

To understand the relationship between corporate and urban systems in the intangible asset accumulation process, two distinctive and contrasting global firms are examined: Samsung Electronics Corporation (SEC) and Lutronic Corporation. As globalization and knowledge-based society are the major global megatrends, we selected these case studies on the basis of knowledge-intensive sectors, global firms, and the importance of intangible assets. Firm size was also considered for comparison. As one of the leading global firms in the electronics sector, SEC understands the increasing importance of intangible assets for firm growth. Lutronic is one of the leading small firms in the field of laser medical appliances and a global actor in terms of market and investments. SEC and Lutronic are knowledge-intensive firms, and both show high R&D investment to total sales ratios (Table 1).

The two firms were selected because these firms exemplify the important role of intangible asset accumulation in their competitiveness. Both are leading global actors despite the difference in size. SEC began producing electronic products in 1969. In the last four decades, SEC has evolved from a manufacturer of low-priced products to a leading global firm producing high-value-added products. SEC has also transitioned from a subcontractor to a flagship innovator that produces new products. Currently, SEC is shifting its focus from innovation to the integration of innovation, marketing, and culture. By contrast, Lutronic was established in 1997 as a small high-tech firm and was known as a venture firm when Korea was confronted with a financial crisis. During the last 16 years, Lutronic has evolved into a leading innovator of advanced aesthetic and medical lasers and related technology, with investments in the United States and Japan. Table 1 presents general corporate information on SEC and Lutronic.

The operating units of the two firms were historically analyzed using various materials, such as corporate annual reports, newspaper materials, web materials, and statistics related to the firms' spatial and structural growth strategies. Five in-depth interviews with firm leaders were conducted in January and October 2012 to clarify the interaction and the accumulation processes. Interviews focused on the importance of intangible knowledge, the source and network of information and knowledge, and the significance of regions in

Table 1. Corporate profiles of SEC and Lutronic.

SEC
Year established: 1969
Evolutionary stages:
1969–1971: Founding period
1980s: Export and foreign-oriented development
1990s: Advance for the new century
2000–2003: Leading digital convergence innovation
2004 ~ present: Building base of Super First-Class Firm
Sales and profits:
2011 Total sales: About US$140billion
2011 Total profit: About US$14billion
R&D expenditure and others:
About 6% of total sales: 10th in the world (2009)
USA patent registration number: 2nd following IBM (2010)
Lutronic
Year established: 1997
Evolutionary stages:
1997–1999: Founding period
2002 ~ present: focusing export
2006: KOSDAC listed
2007 ~ present: Foreign subsidiaries and R&D center
Sales and profits:
2011 total sales: About US$45 million
Continuous increase of sales since 1997
Continuous + profit since 2000
R&D expenditure and others:
16–17% of total sales
10th in the aesthetic devise in the world
3rd world market share in skin rejuvenation devise

Source: Gathered from the internal materials of SEC and Lutronic, websites, interview surveys, and newspapers.

the accumulation of knowledge and intangible assets. The SEC interviews were conducted with the firm's former president, as well as current executive director and public relations director. The Lutronic interviews were conducted with the firm's president and executive director.

To understand the relationship between corporate and urban systems, we first present the results of the examination of the evolution of the spatial corporate system and intangible asset accumulation through global networks and localized processes over time. Subsequently, we analyze intangible asset accumulation through global networks and localized processes and the relationship of firms with urban systems. The three hypotheses are confirmed qualitatively based on the results of the analysis of corporate data as well as the interviews.

Evolution of spatial corporate systems

SEC has grown remarkably since its establishment in 1969. Within 10 years, SEC has managed to become a multinational corporation with a wide array of foreign investments. Located in the capital region of Korea, SEC started with the production of black and white TVs in 1970, after which it expanded its product lines to include refrigerators, air conditioning units, washers, and color TVs. Since the 1980s, SEC has diversified to high-tech products such as semiconductors, personal computers, mobile phones, tablets, and automobiles. SEC has been involved in FDIs since its first investment in the United States in 1978. For 10 years beginning in 1978, SEC invested in foreign subsidiaries such as branch offices in countries of the Global North, including the United States, Germany, the United Kingdom, Australia, and Canada. The decade of the 1980s is considered the key period for SEC's foreign subsidiary investments. Since 1982, SEC has invested in manufacturing plants and sales offices, with dramatic expansion in the 1990s. The firm's manufacturing investments have been mainly greenfield enterprises. The production investments began in countries of the Global North (such as the United States and the United Kingdom) and were extended to industrializing Global South countries such as Thailand, Malaysia, India, and Brazil (Table 2). The foreign R&D investments of SEC were started in 1987, with the late 1980s and 2000s as the key periods. These investments were also initiated in the Global North and eventually extended to newly industrializing economies (Table 2).

Lutronic started exporting its products, four years after its establishment, to Asia with subsequent extension to North America and Europe. Lutronic pursued foreign investments beginning with a subsidiary investment in the United States in 2007 (Table 2). In 2010, the firm invested in an R&D center in the United States. Hence, despite its small size, Lutronic has already engaged in global activities and network expansion to seek knowledge and information. Thus far, the company's foreign investments are primarily in the Global North, but there are plans for expansion into mainland China.

The evolution of the spatial corporate system of SEC suggests that corporate systems extend successively over time from foreign subsidiaries to manufacturing plants, sales offices, and R&D centers—as well as from advanced countries to industrializing countries. Lutronic is considerably smaller than SEC in terms of sales and production as it is a relatively younger company. However, Lutronic has become one of the leading global firms in the skin rejuvenation device industry, a result of high levels of R&D expenditure. In fact, the Lutronic CEO stated, "we want to continue R&D activities and technological development through keeping more than 15% of total sales allocated to R&D

Table 2. Evolution of spatial corporate system through global networks and localized processes over time.

SEC

- Foreign local subsidiary (1978–1988): Key period→1980s:
  Advanced countries: USA, Germany, UK, Australia, Canada
- Foreign plant investment (1982–2011): Key period→1990s:
  From Advanced to Industrializing Countries: Portugal, USA, UK; Thailand, Mexico, Malaysia, Brazil, Slovacia, India, Russia
- Foreign R&D Center (1987–2004): Key periods→late 1980s and 2000s:
  From Advanced to Industrializing Countries: USA, Japan, UK, Russia, Israel, China, India, Poland
- Foreign industrial complex (1995;1996): Key period→1990s: UK, Mexico

Lutronic

- 2001: Export to Asia
- 2004: Export to NA and Europe
- 2006: Listing KOSDAK (high tech venture firm)
- 2007: USA Corporation
- 2008: Japan Corporation
- 2010: Global R&D Center

Source: Gathered from the internal materials of SEC and Lutronic, websites, interviews, and newspapers.

expenditure." Unlike SEC, Lutronic has started exporting to Asia because penetrating the Asian aesthetic device market is relatively easy, given the similarities in the skin characteristics of Asians (i.e., skin sensitivity) with those of Koreans. Our interviewees emphasized that such characteristics are different from those of Western people.[1] Overall, the contrast between the case firms suggests that spatial corporate systems evolve from subsidiary to R&D and are extended from advanced countries to industrializing countries.

Intangible asset accumulation through global networks and localized processes over time

Along with the evolution of spatial corporate systems, the accumulation of intangible assets in the form of knowledge, technical know-how, marketing, brand recognition, and so on by SEC and Lutronic has been continuous. A firm can employ several ways to accumulate intangible assets. For the two case study firms, the most important mechanisms include formal and informal networks within local area clusters, continuous R&D investments, technology cooperation, strategic alliances, and short-term activities such as organizing and participating in temporary clusters (Table 3). A recent study has suggested that the mobility of prominent scientists in the past has simultaneously increased knowledge flow and the creation of localized intangible assets (Schiller & Revilla Diez, 2012), suggesting that FDIs in the global economy influence the accumulation of regional intangible assets.

Formal and informal networks within the local areas of foreign countries, which are formed by individual corporate units, are critical for a firm's intangible asset accumulation. The accessibility of tangibles and intangibles is considered in the initiation of

Table 3. Accumulation of intangible assets through the global networks and localized processes over time.

SEC

- Development of World New Products (1990~): Key periods: 1999; 2004–2006; 2009–2010
- Technology cooperation (1999~): Key periods: 2007; 2010
- Technology strategic alliances (1995~): Key period: 2010~2011
- M&As (2010~): Key period: 2010~2011
- Supporting sports team, Expo and Trade Show (2005~): Key period: 2010 2011
- Continuous R&D investments (Since 1980s): Key period: Since late 1990s
- Investing to new industry (2009~): Key period: 2010~: Medical sector: bio-similar, medical devise; health research

Lutronic

Important sources of knowledge and information for innovation:
- External network and informal meeting of CEO
- R&D; marketing; trade fair and exhibition

Important regions as sources of knowledge and information for innovation:
- Korea: Interactions in the Capital Region for all kinds of economic actors
- USA, EU: for half of the sources (Univ., Gov., research manpower, customer, supplier)
- Japan: for informal meeting of research manpower, customer

Major partner of cooperative network
- University; Customer firm; Competitive firms
- All are in Seoul

Source: Gathered from the internal materials of SEC and Lutronic, websites, interviews, and newspapers.

greenfield manufacturing investments. As indicated in the interviews with the two firms, the subsidiary and manufacturing investments made by both companies began with industrialized countries of the Global North because of the firms' goals to gather market information and technical knowledge and because of the easy access to input factors. SEC's R&D investments in innovation clusters in prominent Global North regions—such as Silicon Valley—were initially aimed at exploring knowledge and technology in the local cluster and collaborating with local specialists and firms. This objective demonstrates the behavior of exploring intangible assets accumulated in local clusters. In the early 1990s, SEC attempted to maintain close local networks to acquire knowledge and technology, as well as to cooperate with Silicon Valley firms in the early stage of its R&D investments (Park, 1994, 1996). On the one hand, SEC practiced continuous and intensive R&D investments. On the other hand, the company maintained technical cooperation and strategic alliances with foreign firms in advanced countries to continuously accumulate and explore new knowledge and technologies required for new product development. The asset-seeking and asset-exploring behaviors of SEC are believed to have contributed to its continuous development of new products since the 1990s. Recently, SEC aggressively organized and participated in temporary clusters, such as expos and trade fairs, to enhance its innovative networks. Temporary clusters have significantly affected the acquisition of new knowledge and product information and the establishment of innovative networks (Bathelt & Glückler, 2011; Rinallo & Golfetto, 2011; Torre, 2008). Trade shows support the development of marketing capabilities into complementary assets, which are important in ensuring effective innovation and competitive advantage (Rinallo & Golfetto, 2011). SEC's R&D investments suggest that clusters and agglomeration are important at the local level, whereas global cluster networks and temporary clusters are important in seeking intangible assets (H2).

Sales network centers are outposts for exploiting marketing bundles of intangible assets in foreign countries. SEC's interactions with other regions and markets within these countries are good opportunities to accumulate marketing information and knowledge. SEC has expanded its sales network centers globally to seek new information and knowledge and exploit global marketing information. Local networks in production centers contribute to savings in transaction costs and are helpful in accumulating intangible assets to be transmitted to regional headquarters. The interview results suggest that it is quite difficult to differentiate seeking from exploiting behaviors. In recent years, the operating units of different firms have demonstrated asset-seeking and asset-exploiting behaviors simultaneously in one place as well in different places (H3).

In the case of Lutronic, the formal and informal relations of the CEO have served as important sources of knowledge and information for innovation. Lutronic's CEO has a strong external network with professors of medical schools, scientists connected with research labs, and engineers of high-tech firms—all of whom are excellent sources of knowledge and information on new products and processes. In addition to conducting internal R&D activities, the CEO also explores new knowledge and enhances the Lutronic brand by periodically visiting institutions and initiating memoranda of understanding. The CEO visits medical schools and hospitals in the United States and Europe several times a year and explores knowledge related to aesthetic and medical laser technology and devices. Participation in temporary clusters serves as an important resource in the acquisition of intangible assets as well. Staff members in the R&D sector participate in international scientific conferences, workshops, and meetings, where they present research articles and explore new knowledge.

Additional studies must be conducted in temporary clusters to fully understand how knowledge creation, diffusion, and innovation may take place beyond regional boundaries. Nevertheless, trade fairs are good opportunities for small firms such as Lutronic to gain access to information on effective marketing and new products. Thus, Lutronic periodically participates in trade shows and conferences to establish new knowledge networks and strengthen its existing networks with other scientists and firms. The company also participates in trade shows and conferences to gather marketing information and investigate consumer behavior. Accordingly, Lutronic considers asset-seeking and asset-exploiting behaviors as simultaneous actions, although the strength of each behavior might vary from country to country.

On the one hand, Lutronic seeks intangible assets by closely interacting with aesthetic service clinics and hospitals in the capital region of Korea. On the other hand, the company seeks knowledge and information for innovation through formal and informal networks with universities, governments, and research centers of advanced countries. Networks with customers and suppliers in the high-tech clusters of Global North countries are also important sources of improvement of new products and new information on production processes. Lutronic has designated major partners including universities, customers, and supplier firms in Seoul, which led to the establishment of cooperative networks. In addition, Lutronic promotes marketing activities in the emerging markets of Asia and Russia to exploit intangible assets that are related to product and market information. Considering that Lutronic is a small firm, the role of its CEO in acquiring intangible assets, in addition to the firm's formal and informal networks with major economic players, is critical.

Lutronic clearly maintains a close innovation network in the capital region of Korea and strong networks with major high-tech clusters in the United States. The firm also participates actively in international trade shows, conferences, and workshops. Therefore, the interview results of Lutronic support H2 and H3. For intangible assets, agglomeration and clusters are important at the local level, whereas transnational cluster networks and temporary clusters are important at the global level (H2). The interview results also confirm that Lutronic seeks and exploits intangible assets simultaneously (H3).

Interactions with urban systems

Main headquarters, regional headquarters, R&D centers, production plants, and sales offices can be regarded as the major units of corporate systems. To relate corporate systems with urban systems, cities are classified based on the country and the city levels within the said country. Following the World Bank classification system, we classified the countries as follows: advanced countries (A), industrializing countries (B), and less industrialized countries (C). Cities are classified as the capital or the largest metropolitan area, including suburbs (A); large metropolises with a population of more than one million, including suburbs (B); and medium and small cities (C) (Table 4).

As for the two firms in this study, their main headquarters are located in the capital region of Korea to ensure easy communication with and control of diverse corporate units, access to the national government, close relations with financial firms, and close networks with leading universities and research labs. Moreover, regional headquarters are located in the major capital regions of the divided global markets according to strategic considerations of control and management functions. SEC divides the global market into eight regions, as seen in the regional headquarters set-up in Table 4. The regional headquarters are located in the capital region or in the largest metropolis of a leading country within each of the divisions of the global market (Table 4).

Table 4. Corporate system and urban system.

SEC Regional HQs (4AA, 6BA) – North America: Ridgefield Park, NJ, USA: AA – Europe: Surrey, UK: AA – SE Asia: Singapore, Singapore: AA – China: Beijing, China: BA – Japan: Tokyo, Japan: AA – CIS & Baltics: Moscow, Russia: BA – Latin America: Sao Paulo, Brazil: BA – SW Asia: New Delhi, India: BA – Middle East & Africa: Dubai: BA – Africa: Johanesburg, South Africa:BA	Production Centers (1AC, 2BA, 20BB, 7BC, 2CA) – North America: 1AC,1BB – Europe: 1BA, 2BC – SE Asia: 3BB, 2BC, 2CA – China: 12BB, 1BC – CIS and Baltics: 1BC – Latin America: 1BA 2BB – SW Asia: 2BB – Middle East & Africa: 1BC
R&D Centers (2AA, 3AB, 3BA, 4BB) – North America: AB (San Hose); AB (Dallas); – Europe: AA(London); BA(Warsaw) – SE Asia and Japan: AB (Yokohama) – China: BA (Beijing); BB (Suzhou); BB (Nangyung) – CIS and Baltics: BA (Moscow) – Latin America: – SW Asia: BB (Bangalo); BB (Delhi) – Middle East & Africa: AA (Tel Aviv)	Sales Network Centers (14AA, 3AB, 29BA, 8CA) – North America: 2AA, 2AB, 1BA – Europe: 6AA, 1AB, 9BA – SE Asia, Japan & Australia: 4AA, 6BA – China and HK: 1AA,1BA – CIS and Baltics: 2BA, 1CA – Latin America: 1AA, 4BA, 2CA – SW Asia: 1BA – Middle East & Africa: 5BA, 5CA
Lutronic HQ: Capital Region (AA) Regional HQ: New York (AA), Tokyo (AA) R&D Center: San Jose (AB) Sales Office: Birmingham (AB), Beijing (BA)	Export to 60 countries Extending from Asia to Western countries – Export share of North America, EU, Japan: 22% in 2002 → 50% in 2010 4 Market differentiation for aesthetic device by skin color – Asia; North America; Europe; Latin America – China and Russia: Emerging Market

Note: Corporate System: Regional HQ, R&D, Production, and Sales.

Urban System: Considering both country and city level

Country Level (First)	City Level (Second)
A: Advanced country	A: Capital/largest metropolitan area including suburbs
B: Industrializing country	B: Large metropolis(+Million) including suburbs
C: Less industrialized country	C: Medium/small city

Source: Gathered from the internal materials of SEC and Lutronic, websites, interviews, and newspapers.

The locations of R&D centers show considerably diverse patterns. Specifically, the locations match either capital regions or large metropolises, where innovative clusters or regional innovation systems are well developed. Each division of the global market has at least one R&D center. Examples of the so-called capital regions are London, Beijing, Moscow, and Tel Aviv, where high-tech clusters or clusters of knowledge-intensive services are well developed. Noncapital metropolises, such as San Jose, Dallas, Warsaw, Yokohama, Suzhou, Nangyung, Bangalo, and Delhi, are major innovation clusters or high-tech clusters within the respective divisions of the global market (Table 4).

Most production centers are located in industrializing countries, in contrast to the pattern of regional headquarters. For example, only one plant is located in an advanced country, but in a small or medium city. Two plants are located in a less industrialized country, but situated in a capital region. Most foreign plants (more than three quarters) are located in the large metropolises of industrializing countries. Almost half of production centers are located in China, whereas no production center is located in the regional divisions of the Middle East or Africa. This observation suggests that production centers are closely related to industrial clusters and actual markets. Production centers are mainly located in cities where industrial clusters are well developed to enhance the efficiency of the production networks in local areas. The interview results suggest that for investments in production centers, the savings in transaction costs contributed by the production networks of industrial clusters within industrializing countries are considered, and intangible assets are exploited in relation to production activities. Given the progress of the so-called information society, modern production centers at present have long-distance networks for transactions, although these networks are focused on selected areas, such as industrial clusters or innovation clusters (Park, 2004).

Sales network centers show a highly diverse pattern that is considerably different from that of other corporate system units. Sales network centers are dispersed throughout the world, from advanced countries to less industrialized countries. A considerable number of sales network centers are located in the less industrialized countries of the Middle East and Africa, and Latin America, as well as Baltic countries and those that belong to CIS. In this case, the centers are located in either capital regions or large metropolitan areas. In general, approximately 94% of sales network centers are located in capital regions. Different functions are expected from the location of the centers based on the hierarchy of urban systems in the global economic spaces. First, sales network centers located in the capital regions of advanced countries seek various types of intangible assets in addition to marketing-related knowledge and product information. Second, sales network centers in the capital regions or large metropolises of industrializing countries pursue both asset-seeking and asset-exploiting activities simultaneously. Third, sales network centers in less industrialized countries gather information on local resources and suppliers to exploit intangible assets for the future development of the sales market and production activities beyond direct sales activities.

The location of operating units in the global space supports the first hypothesis, which supposes that the hierarchy of a corporate system is related to the hierarchy of an urban system in the flow of intangible assets within the corporate system (H1). The results of the interviews confirm that SEC primarily considers the major capital regions in a continent when selecting locations for its regional headquarters to control operating units in the region, explore knowledge and information, find innovative environments for R&D centers, identify markets for sales network centers, and access input materials and markets for production centers. Such consideration of different locations based on function confirms that high-tech MNFs consider the correlation between the external environments of cities and different functions in gathering intangible assets in inter-organizational networks.

Lutronic only has two regional headquarters, which are located in Tokyo and New York. These headquarters represent the firm's control centers in Asia and the West, respectively. The firm considers control and communication network functions in identifying the location of their regional headquarters in Eastern and Western countries to enhance the effectiveness of the acquisition of diverse information on skin characteristics, medical information, and information on new technology. The R&D center is located in Silicon Valley, the global innovation cluster, for effective seeking and accumulating of intangible assets such as knowledge, marketing, and new product information. The sales offices are located in the capital region of China (Beijing) and in a medium-sized US metropolis (Birmingham). Although the operating units of Lutronic are few, the case partially supports the first hypothesis (H1). However, H1 is not fully supported because of Lutronic's lack of operating units.

Lutronic broadly identifies four global markets—Asia, North America, Europe, and Latin America—and pursues different strategies by region. In the interview, the representatives of Lutronic identify the levels of differences in term of skin characteristics, such as sensitivity and color, in the four global markets. Thus, they pursue different strategies in marketing and technical development. In recent years, China and Russia have been considered as emerging markets. In one sense, Lutronic has focused on asset-exploiting and asset-seeking activities in emerging and advanced markets, respectively. In the process, however, Lutronic simultaneously pursues asset-seeking and asset-exploiting behaviors despite variations in the emphasis of different economies. Accordingly, the relationship between corporate and urban systems in terms of seeking and exploiting intangible assets can be clarified further when Lutronic expands to a global market and establishes additional operating units in the future.

The results of the analysis of the two case study firms suggest the existence of a strong relationship between corporate and urban systems in global economic spaces. Interactions between firms and regions are based primarily on the hierarchies of corporate and urban systems. The results of the in-depth interviews and data analysis of the location of corporate system units show that firms accumulate intangible assets through interactions with urban systems in global economic spaces. Thus, the case study generally confirms the conceptual framework and the three hypotheses and facilitates the identification of the processes of intangible asset accumulation in terms of asset-seeking and asset-exploiting activities.

Concluding remarks

This study developed a conceptual framework of the relationship between corporate and urban systems in intangible asset accumulation, with the aim of extending Wheeler's work on such systems in relation to global spaces and intangible assets. Two distinctive and contrasting global firms were historically analyzed using various materials and statistics on both firms' spatial and structural growth strategies. In-depth interviews were conducted to clarify the interactions and accumulation processes within each firm.

The results of the interview surveys and the analysis of the operating unit locations of the two cases support the three hypotheses. However, this study did not test the three hypotheses thoroughly because the examination was focused on the overall trends in the relationship between corporate systems and urban systems with regard to intangible assets.

The data analysis and the interview surveys suggest that firms' intangible assets are critical in ensuring their competitiveness within a knowledge-based information society. The intangible assets of firms can be increased through the accumulation of knowledge, technology, human capital, brand values, and so on. Such accumulation is possible through formal and informal interactions and networks with suppliers, customers, competitors, governments, and communities. In the process of intangible asset accumulation and interactions with urban systems, firms differentiate their strategies for cities based on the country's development level and the status of the urban system. Intangible assets are also important in maintaining the competitive advantages of cities and regions.

Accumulation of local intangibles is the result of the evolutionary development of the relationships between the corporate and urban systems.

The accumulation of intangible assets in a given region enables industrializing countries to attract FDIs. The externalities that stem from knowledge spillover from firms can influence the accumulation of regional intangible assets and result in a positive relationship between firms and regions in a knowledge-based economy (Artis et al., 2012). However, policies devoted to the generation of regional intangible assets should consider the specificities of regions in relation to urban systems. This suggestion is based on the results of this study that show that the acceleration of intangible asset accumulation and the effects of intangible assets are not homogenous over hierarchies of urban systems.

The conceptual framework and case studies presented in this work have limitations. Nevertheless, we believe that this study can contribute to the reexamination of broad issues in the relationship between corporate systems and urban systems, particularly with regard to intangible asset accumulation in a knowledge-based information society. The results from such reexamination are important in understanding dynamic global economic spaces, an area in the field of geography which has been relatively neglected.

Acknowledgment

The author is grateful to the editors and three anonymous referees for very helpful comments and suggestions.

Funding

This work was supported by the National Research Foundation of Korea [grant number 2008-B00027].

Note

1. According to our interview respondents in Lutronic and an aesthetic service firm, the skins of Asians are much more sensitive than those of Westerners. Moreover, interviewees believe that Asians follow a more complex and difficult skin care regimen than Westerners do.

References

Anson, Wester (2007). *The intangible assets handbook: Maximizing value from intangible assets*. Chicago, IL: American Bar Association.

Artis, Michael J., Miguelez, Ernest, & Moreno, Rosina (2012). Agglomeration economies and regional intangible assets: An empirical investigation. *Journal of Economic Geography, 12*(6), 1167–1189.

Asheim, Bjørn T., Boschma, Ron, & Cooke, Philip (2011). Constructing regional advantage: Platform policies based on related variety and differentiated knowledge bases. *Regional Studies*, *45*(7), 893–904.

Bathelt, Harald, & Glückler, Johannes (2011). *The relational economy. Geographies of knowing and learning*. Oxford: Oxford University Press.

Bianchi, Patrizio, & Labory, Sandrine (Eds.). (2004). *The economic importance of intangible assets*. Aldershot: Ashgate.

Birkinshaw, Julian, Brannen, Mary Yoko, & Tung, Rosalie L. (2011). From a distance and generalizable to up close and grounded: Reclaiming a place for qualitative methods in international business research. *Journal of International Business Studies*, *42*(5), 573–581.

Dicken, Peter, & Malmberg, Andres (2001). Firms in territories: A relational perspective. *Economic Geography, 77*(4), 345–363.

Dunning, J. H. (1995). Re-appraising the eclectic paradigm in an age of alliance capitalism. *Journal of International Business Studies*, *26*(3), 461–491.
Fey, Carl F., & Birkinshaw, Julian (2005). External sources of knowledge, governance mode, and R&D performance. *Journal of Management*, *31*(4), 597–621.
Gong, Hongmian, & Wheeler, James O. (2002). The location and suburbanization of business and professional services in the Atlanta area. *Growth and Change*, *33*(3), 341–369.
Hymer, Stephen H. (1976). *The international operation of national firms: A study of direct foreign investment*. Cambridge, MA: MIT Press.
Kramer, Jan-Philipp, Marinelli, Elisabette, Iammarino, Simona, & Revilla Diez, Javier (2011). Intangible assets as drivers of innovation: Empirical evidence on multinational enterprises in German and UK regional systems of innovation. *Technovation*, *31*(9), 447–458.
Kramer, Jan-Philipp, Revilla Diez, Javier, Marnelli, Elisabette, & Iammarino, Simona (2009). *Intangible assets, multinational enterprises and regional innovation in Europe* (Working Paper 1.3.b.). IAREG, 1–69.
Kumar, Nagesh (1998). *Globalization, foreign direct investment and technology transfers: Impacts on and prospects for developing countries*. New York, NY: Routledge.
Lösch, August (1952). *The economics of location* [*Die Räumriche Ordnung der Wirtschaft* (1940)]. (W. Woglom, Trans.). New Haven, CO: Yale University Press.
Makino, Shige, Lau, Chung-Ming, & Yeh, Rhy-Song (2002). Asset-exploitation versus asset-seeking: Implications for location choice of foreign direct investment from newly industrialized economies. *Journal of International Business Studies*, *33*(3), 403–421.
Marrocu, Emanuela, Paci, Raffaele, & Pontis, Marco (2012). Intangible capital and firms' productivity. *Industrial and Corporate Change*, *21*(2), 377–402.
O'Mahony, Mary, & Vecchi, Michela (2009). R&D, knowledge spillovers and company productivity performance. *Research Policy*, *38*(1), 35–44.
Park, Sam Ock (1994). High technology industrial development and formation of new industrial district: Theory and empirical cases. *Journal of Korean Geographical Society*, *29*(2), 117–136. (In Korean with English summary).
Park, Sam Ock (1996). Networks and embeddedness in the dynamic types of new industrial districts. *Progress in Human Geography*, *20*(4), 476–493.
Park, Sam Ock (2004). The impact of business to business electronic commerce on the dynamics of metropolitan spaces. *Urban Geography*, *25*(4), 289–314.
Park, Sam Ock, & Wheeler, James O. (1983). The filtering down process in Georgia: The third stage in the product life cycle. *The Professional Geographer*, *35*(1), 18–31.
Raisch, Sebastian, Birkinshaw, Julian, Probst, Gilbert, & Tushman, Michael L. (2009). Organizational ambidexterity: Balancing exploitation and exploration for sustained performance. *Organization Science*, *20*(4), 685–695.
Regional Studies. (2012). Special issue: Intangible assets and regional economic growth. *Regional Studies*, *46*(10), 1277–1416.
Rinallo, Diego, & Golfetto, Francesca (2011). Exploring the knowledge strategies of temporary cluster organizers: A longitudinal study of the EU fabric industry trade shows (1986–2006). *Economic Geography*, *87*(4), 453–476.
Schiller, Daniel, & Revilla Diez, Javier (2012). The impact of academic mobility on the creation of localized intangible assets. *Regional Studies*, *46*(10), 1319–1332.
Surinach, Jordi, & Moreno, Rosina (2012). Introduction: Intangible assets and regional economic growth. *Regional Studies*, *46*(10), 1277–1281.
Taylor, Michael, & Asheim, Bjørn (2001). The concept of the firm in economic geography. *Economic Geography*, *77*(4), 315–328.
Tommaso, Marco R., Paci, Daniels, & Schweitzer, Stuart O. (2004). Clustering of intangibles. In P. Bianchi & S. Labory (Eds.), *The economic importance of intangible assets* (pp. 73–102). Aldershot: Ashgate.
Torre, A. (2008). On the role played by temporary geographical proximity in knowledge transmission. *Regional Studies*, *42*(6), 869–889.
Weber, Alfred (1929). *Theory of location of industries* [*Über den Standort der Industrien* (1909)]. (C. J. Friedrich, Trans.). Chicago, IL: University of Chicago Press.
Wheeler, James O. (1981). Effects of geographical scale on location decisions in manufacturing: The Atlanta example. *Economic Geography*, *57*(2), 134–145.

Wheeler, James O. (1986). Corporate spatial links with financial institutions: The role of the metropolitan hierarchy. *Annals of the Association of American Geographers*, *76*(2), 262–274.
Wheeler, James O. (1988a). The corporate role of large metropolitan areas in the United States. *Growth and Change*, *19*(2), 75–86.
Wheeler, James O. (1988b). Corporate activities and the urban hierarchy in Georgia. *The Southeastern Geographer*, *28*(2), 97–109.
Wheeler, James O. (1999). Local information links to the national metropolitan hierarchy: The southeastern United States. *Environment and Planning A*, *31*(5), 841–854.

Macau's role as a recreation/tourist center in the Pearl River Delta city-region

Clifton W. Pannell[a] and Philip H. Loughlin[b]

[a]*Department of Geography, University of Georgia, Athens, GA, USA;* [b]*Department of Economics, Maryville University, Cambridge, MA, USA*

The return of Macau to Chinese sovereignty in 1999 brought far-reaching change to Macau as a special administrative region (SAR). China's policy and plans for a revitalized Macau as a key urban node offers an opportunity to conceptualize a leading role for Macau as the pre-eminent recreational and tourist center in the dynamic and globally connected Pearl River Delta city-region. We examine and assess the new policies that have led to Las Vegas-style casinos bringing explosive growth in visitors, gambling revenues, and Macau's economy. New problems of corruption, crime, and social pathologies have appeared as well owing in part to the secretive VIP gambling centers for high rollers from the mainland. Has Macau met China's goals and expectations for a great tourist and recreational center? Our findings indicate a mixed record with more challenges emerging in the wake of China's recent anti-corruption campaign.

Introduction

Macau, a former Portuguese enclave and now a tiny special administrative region (SAR) on the western flank of the Pearl River Delta (PRD) across from Hong Kong, is a most unusual and intriguing city.[1] Its checkered history as an early colonial trading post and center for religious proselytizing was followed by commercial stagnation once the port of Hong Kong captured much of the China trade in the mid-nineteenth century. Macau required new sources of revenue, and this led to more nefarious practices. These included expanding the crude gambling activities, prostitution, and the malevolent coolie trade, corralling human beings and shipping them abroad for indentured labor services (Veeck, Pannell, Smith, & Huang, 2011, pp. 328–341).

Macau has long sought an avenue for prosperity, but for many years simply served as a seedy backwater for residents and visitors from Hong Kong who were looking for more dissolute and salacious activities not allowed in the British colony. Gambling, especially as it evolved into a larger enterprise with associated casinos and hotels, increasingly became Macau's answer and the most secure and reliable source of revenue in the second half of the twentieth century. Regrettably, gambling also attracted criminal elements and triads who saw a niche that could be exploited with little threat of reprisals from local law enforcement (Gaming, Inspection, and Coordination Bureau, Macao SAR, 2015).

The handover to China in December 1999 brought far-reaching change to Macau. First, the Chinese government quickly worked to end the overt illegal and violent activities associated with the local criminal elements. A new era of modern gambling began in which the term "gaming" was substituted, and the gambling monopoly was ended. New casino entrepreneurs from North America were allowed to enter the market. These new enterprises introduced new and modern approaches that sought to turn gaming into a key attraction to advance the tourist industry (Hing, 2005). A restructured Macau, featuring top-quality hotel resort centers focused around elegant casinos, was then given a special role as the key recreation/tourist center in the rapidly developing city-region of the PRD (Figure 1) (Gaming, Inspection and Coordination Bureau, Macao SAR, 2015).

In this paper, we seek to answer several key research questions inspired from Macau's remarkable renaissance and recent development: (1) What has been the economic impact of gaming liberalization on tourism in Macau? (2) What has been the social impact of gaming liberalization in Macau? (3) How have the economic and social impacts of Macau's gaming liberalization affected the goal of making Macau an attractive tourist and leisure center? From these three research questions, we derive a larger issue as to the manner in which Macau is, or is not, meeting the expectations of China's policy makers and planners as expressed in the 2009 comprehensive plan for the PRD city-region? More specifically, is Macau in fact evolving into a world-class tourist/leisure center and fulfilling its desired urban function as a third anchor leg of the PRD city-region? This would offer evidence of its success in its role in the expanding global outreach of the emerging city-region of the PRD.

The paper will review appropriate literature on Macau and the PRD region as well as economic and social issues related to gaming as a specified activity. Our conceptual framework of the evolving metropolitan region of the PRD builds on the empirical and theoretical work of McGee (1991), Ginsburg, Koppel, and McGee (1991), Yeung and Hu (1992), Lin (1997) and others, while the concept of city-regions within a global context will be examined through the theoretical and analytical writings of Scott (2001) and Dicken (2007). In addition, we will briefly consider the remarkable recent impacts that the anti-corruption drive in China has brought to Macau, which create a good opportunity to review the impact of central state to local region, a link to the relational and territorial geographies of urban policy. McCann and Ward (2011) have provided a thoughtful conceptualization of these relationships. Data and statistics employed were obtained from local agencies in the PRD and Macau and will be used to support our interpretations and conclusions. Such data, information, and analysis will relate to the above research questions, and we will seek to explain and answer them specifically in our interpretive and concluding remarks.

In examining and analyzing Macau, we have taken two other major tourist cities, one of which is based on gaming, for comparative purposes that add insight into our understanding and explanation of Macau. These are its neighbor, Hong Kong, and its North American gambling model, Las Vegas.

The idea of city-regions: a conceptual framework

The idea of a city-region or conurbation is not new and originated in the conceptual thinking of the French geographer Jean Gottman (1961) in the 1950s in his writings about the urbanized northeast corridor of the United States, a conglomeration of cities linked through a dense transport network, which he called "megalopolis." While his idea has been expanded to describe and explain a variety of urban clusters in many parts of the

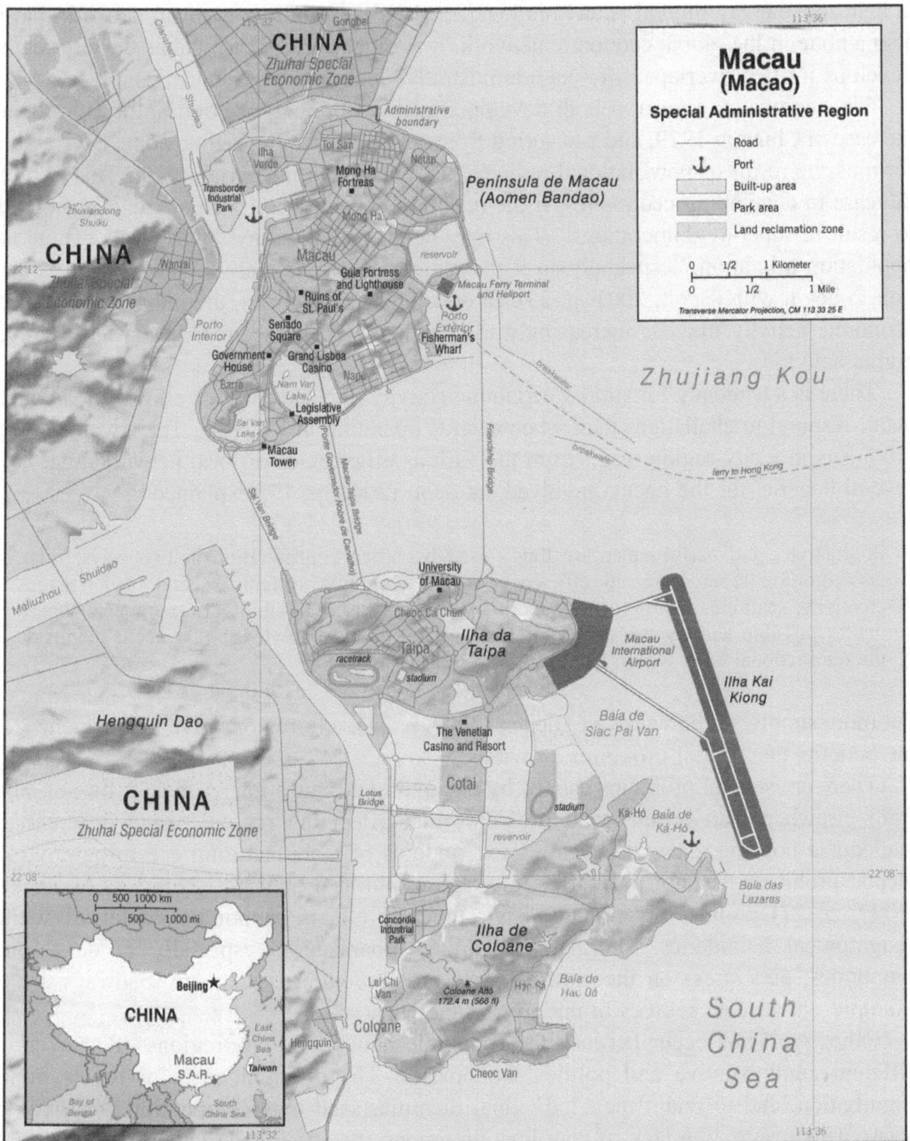

Figure 1. Macau. Source: US Government (2015).

world, our focus is on the identification and analysis of a large urban region, the PRD, and one of its key subunits, Macau, in southeast China (Figure 1).

Before going into elaborate details on the PRD, it is helpful to review the conceptual ideas of Allen J. Scott (2001), Peter Dicken (2007) and others concerning city-regions or regional networks of cities. They provide an innovative way of thinking about urbanization and its spatial expression. These conceptualizations help explain much of what can be observed in the PRD today.

The city-region, as described by Scott, can be a familiar metropolitan area or a hyper-agglomerated urban area. Included in the concept is the surrounding hinterland. The area is held together by mutual economic benefits. It is viewed as a regional network that is also a node in the global economic network. It is often polycentric, and as the city-region develops it often overlaps different administrative and political boundaries.

City-regions may form in both developed and developing economies, the latter being the case in China in 1979, and can spring from several different origins. Some city-regions form as the result of new national policies that encourage industrialization. This is often the case in developing economies. If the resulting development is rapid enough, "it tends to result in super-agglomerations." This may be accompanied by "high levels of internal population migration," especially so if the region had a relatively low level of urbanization to begin with (Scott, 2001, p. 23). In developing economies, the indigenous primary economic activity may be increasingly marginalized by the secondary and tertiary economic activities.

There is a tendency for similar economic activity to cluster, which allows all agents to better respond to challenges than when each is operating in isolation. The advantages for producers in a city-region result from the various efficiencies of local networks that have several benefits for the agents involved, as Scott (2001, p. 17) explained:

> Productivity and performance are thus raised by urban concentration in two ways. First, concentration secures overall efficiency of the economic system. Second, it intensifies creativity, learning and innovation both by the increased flexibility of producers that it makes possible and by the enormous flows of ideas and knowledge that occur alongside the transactional links within localized industrial networks.

Put more simply, in the words of Dicken (2007, p. 21), city-regions give to the agents "... the benefits that spatial proximity provides."

There are several problems caused by expanding urbanization. A huge influx of labor from elsewhere can bring social problems that affect the region—wage differentials, inadequate housing, etc. The negative externalities of manufacturing can bring environmental problems that not only affect the region, but also may go beyond it. As Dicken (2007, p. 27) pointed out, "many environmental effects cannot be contained within geographical boundaries." Further, expanding urbanization, especially in developing economies, puts stress on the hard and soft infrastructure. Inadequate roadways are an example. Inadequate sources of energy are another.

Other problems occur because the economic network in city-regions often overlaps different administrative and political boundaries. The different cities or towns in the conurbation tend to retain their individual identities, and thus competition and rivalries occur. This results in a lack of coordination among the administrative and political units that are part of the network of economic activity. According to Scott (2001, p. 26) "many of the current attempts in developing countries to deal with the problems of their city-regions are based on policy frameworks that are likely to be inadequate."

The solution to these many problems is "The creation of new and responsive frameworks of regional governance capable of sustaining economic development, instigating a sense of cooperative regional identity, and promoting innovative ways of achieving regional democracy and economic fair play is the great challenge for the future" (Scott, 2001, p. 21).

The concept of city-regions in the regional context of the PRD

McGee (1991), Ginsburg et al. (1991), Oshima (1987), and Lin (1997) took this idea of megalopolis or great city-regions and applied and modified it in the context of densely populated, rapidly urbanizing lowland areas of East and Southeast Asia including the great floodplain regions of Japan, Java, and southeastern China. What they postulated was a structural shift in the economy, as increased commercialization of the agricultural economy led to off-farm movement of rural workers to nearby urban areas in the growing industries and increased wages and demand for manufactured products. This in turn led to improvements in the local transport networks and higher personal mobility of labor and consumers (Hsu, Pannell, & Wheeler, 1980). Towns and cities grew, and the surrounding rural areas were increasingly interconnected with the broadening urban nexus in an upward spiral of economic and regional development. A hallmark of this distinctive Asian urbanization was the extraordinarily high densities of rural and urban populations and the number and size of the cities. Nowhere is this seen in greater numbers than in China.

The PRD was characterized as being "one of (China's) most backward regions" (Enright, Scott, & Chang, 2005, p. 13) just a few years before it began its 1979 transformation into a city-region, one of the world's major and most conspicuous manufacturing areas, and the supplier of large quantities of goods for the global market. The immediate cause of the transformation of the PRD was China's Open Door Policy of 1979. This was the national policy origin of the PRD city-region as conceptualized by the city-region theorists. The purpose of the Open Door Policy in part was to attract foreign investment, promote foreign trade, and bring advanced technologies and managerial skills to China (Li & Li, 1999, p. 8). This policy started the period of export-driven economic development in the PRD and marked China's departure from the import-substitution model that had prevailed before. This highly successful new model upon which the PRD was developed became a precursor to the model used later in Macau to develop gaming-related tourism.

The Open Door Policy was first introduced on a limited scale, and the PRD was the policy's major beneficiary. The policy contained a regional strategy that gave priority to coastal areas in southern China, with special attention to Guangdong Province. It was an economic experiment on the part of China, and Guangdong plus neighboring Fujian Province were selected as sites for the experiment in large part because they were of "relatively little financial importance ..." (Ma, 2000, p. 136). In addition, Guangdong had ample space and excess labor to offer and was joined to Macau and Hong Kong, which gave China a conduit to outside resources and markets. China was thus able to make use of this land and excess labor while using foreign capital and technology (Chow, 2002, p. 295). The policy also empowered the local and regional governments of Guangdong by delegating many of the economic decision-making powers.

China's then-paramount leader Deng Xiaoping provided the leadership during the early reform period to allow the Open Door Policy to flourish. His southern trips to the PRD in the early 1990s placed his imprimatur on the legitimacy of economic growth and the use of market incentives to create new wealth in this region of China (Vogel, 2011). Rapid rural-to-urban migration and interior to coastal movement of temporary migrants accelerated the growth of cities and towns in the PRD in parallel with the fast economic growth. Nowhere in the world has the rate and dimension of China's urban transition as seen in the PRD and other major urban mega-regions of China ever been experienced (Chan, 2012).

To implement the policy, four special economic zones (SEZs) were established in Guangdong and Fujian Provinces, including one, Shenzhen, adjacent to Hong Kong and one, Zhuhai, adjacent to Macau. Both SEZs became significant centers of economic activity in the PRD. SEZs were intended to attract foreign investment because of preferential tax and import duty treatment and the ability to repatriate profits and capital (Enright et al., 2005, p. 36). Soon, the PRD saw the emergence of "hundreds of thousands of factories ... many producing for export markets" (Enright et al., 2005, p. 11). As economic activity accelerated in the PRD after 1979, towns and small cities began to cluster around 11 principal cities that were linked together economically if not politically.[2] And thus the PRD became an extreme example of a city-region, a "polycentric or multi-clustered agglomeration ..." (Scott, 2001, p. 18).

One of the distinguishing features of the PRD became foreign investment provided to manufacture goods for export. At first production focused on the domestic market to meet "pent-up demand for a variety of consumer products"; later, manufacturing for export began to advance in the late 1980s (Enright et al., 2005, pp. 38–39). Local firms in the PRD still actively produce for the China market, but foreign investment especially is involved in export manufacturing (Enright et al., 2005, p. 168). As a result, the PRD is often cited as a prime example of export-driven economic development (Lin, 1997). Foreign investment came largely from or through Macau, Hong Kong, and Taiwan, indicating that there was substantial Chinese capital located outside of China eager for the opportunity to take advantage of the new terms in the PRD. The incentives were a large China market of over one billion people and the cost-effectiveness of producing in the PRD for international markets. The low costs, a result of tax advantages and inexpensive land and labor, allowed foreign investors to produce in the PRD and be price-competitive in markets all over the world after transportation costs.

Political complications

As the economy of the PRD developed, one of the major hindrances became the PRD's multiple and often competing political jurisdictions, as explained by Scott (2001). To complicate matters, the "one country, two systems" arrangement under which Macau and Hong Kong were returned to China created boundaries that still retain some of the characteristics of an international border.

Competition among the various political units, including those in Hong Kong and Macau, has resulted in an inefficient use of resources such as underdeveloped areas in some regions of the PRD and redundant facilities in others (Loughlin & Pannell, 2010b). The situation has been characterized as "... the complex and dynamic power struggle in the evolution of the cross-boundary integration of the Greater PRD in the unique and evolving context of 'one country, two systems'" (Yang, 2006, p. 817). What is needed in the PRD is a framework to better coordinate development in the future than in the past.

To address the political complications, meetings have taken place and agreements have been made that move toward a coordinated PRD. A recent document having to do with a comprehensive plan for the PRD, and the focus of this paper, is *Building Coordinated and Sustainable World-class City-region. Planning Study on the Co-ordinated Development of the Greater Pearl River Delta Townships* published in October 2009 (*Planning Study*).[3] This document is a thorough presentation of currently recommended joint strategies of Macau SAR, Hong Kong SAR, and Guangdong Province to work together to achieve a more coordinated development of the city-region in the future. The scope of the recommendations includes spatial organization, transportation, environmental protection, and tourism.

"The major objective of the study is to formulate strategies, action plans and mechanism for coordination in spatial development of the Greater PRD City-Region under the 'One Country, Two Systems' framework ..." (Guangdong Province, Hong Kong SAR, & Macau SAR, 2009, p. 3). The study also stated that "Guangdong, Hong Kong and Macao should jointly build a world-class city-region with coordinated development, sustainability and world competitiveness" (Guangdong Province, Hong Kong SAR, & Macau SAR, 2009, p. 38). The *Planning Study*'s importance is that it "is the first strategic study undertaken at the agreement among the Hong Kong and Macao Affairs Office of the State Council and the governments of Guangdong Province, Macau SAR and Hong Kong SAR" (Guangdong Province, Hong Kong SAR, & Macau SAR, 2009, p. 1), and it recognizes that the effectiveness of a city-region competing in a global economy "depends on whether the concerned cities are administered under a coordinated framework of the city-region" (Guangdong Province, Hong Kong SAR, & Macau SAR, 2009, p. 2).

Under the plan, Macau has a multifaceted role that includes coordinating with other areas of the city-region addressing ecological concerns, spatially balancing economic development, and improving the transportation infrastructure. Among other specific objectives,[4] Macau has a major role in the further development of tourism in the PRD. The *Planning Study*, which notes that Macau has "integrated into the global economy through its gaming and tourism industries for a long time" (Guangdong Province, Hong Kong SAR, & Macau SAR, 2009, p. 29), envisions a "Gold Coast" of tourism that extends along the coast of Guangdong Province (Guangdong Province, Hong Kong SAR, & Macau SAR, 2009, p. 20) and in which Macau's role is "to become the most attractive tourism and leisure center and regional trading and commercial platform in the world" (Guangdong Province, Hong Kong SAR, & Macau SAR, 2009, p. 7).

Macau is a very special situation both for its role as the only place in China that allows casino gambling and for its new position as a key tourist/recreational center for the PRD city-region. Does it fit into that conceptual idea Murphy (2012) asserted within "China's Cultural and Creative Economy," whereby the state, through investment intervention, creates new economies based on tourism that accelerate development in specific locales?

Is Macau an example of the exceptional nature of the Chinese economy and its urban manifestations that Lin and Hu (2011) propose in their provocative essay on the nature of China's recent economic growth? They challenge the "conventional wisdom of neo-liberalism in which privatization is described as a prerequisite for efficient economic growth and as a universally applicable mechanism of societal transition." Macau is clearly special and different, and the long arm of Chinese state intervention is everywhere present when we see the sufferance of state permission and investment to allow this remarkable place to exist within the spatial confines of a people's republic.

We note that the Chief Executive of Macau in a recent policy address has indicated the Macau government will form a "Committee for the Establishment of a World Tourism and Leisure Center" and formulate a five-year plan for its implementation, Office of the Chief Executive of the MSAR Government (2015). This appears to be a reaction to new policy directions from President Xi Jinping and the central state in Beijing (Associated Press, 2014; O'Keefe, 2014, 2015). Can these recent shifts in policy, which as we shall see later, may have a dramatic impact on the economy and future development of Macau, be conceptualized in terms of fast policy shift and transfer invoked in the essays on mobile urbanism in McCann and Ward (2011)?

Macau SAR

Relatively speaking, Macau is a small place. It has an area of only 30 square kilometers and a 2014 population of 640,700. Hong Kong, by contrast, has an area of 1,100 square kilometers and a population of 7.2 million. The mainland part of the PRD has more than 40 million residents. What makes Macau important and unique in China is that it is the only place in the People's Republic of China where casino gambling is allowed.[5] Gambling has been outlawed in mainland China since the Communist Revolution of 1949, except for a state-run lottery. The only legalized gambling activities in Hong Kong are lottery, football betting, and pari-mutuel betting on horseracing.

Origins of Portuguese Macau

The Portuguese were in Macau from 1557 to 1999, and for the last 112 years of that period exercised sovereignty over the territory (Loughlin & Pannell, 2010a; Pannell, 2012; Veeck et al., 2011, pp. 328–341). The Portuguese presence began when the Portuguese, with permission from the Chinese, and under Chinese supervision, established a trading post on the Macau peninsula in 1557. After the success of the British in the First Opium War, and the establishment of the British colony of Hong Kong, the Portuguese sought sovereignty over the Macau peninsula as well as the two islands immediately south of the peninsula, Taipa and Coloane. In 1887, China finally transferred authority over the entirety to Portugal. Over the years, Portuguese power in Macau waned, and sovereignty over Macau was returned to China in the handover of 1999—two years after the similar handover of Hong Kong. The Portuguese left behind a cultural heritage that remains visible in Macau's urban landscape.

History of gambling

Gambling has a long history in Macau. From the beginning, in 1557, the Chinese going to Macau to work brought gambling with them, and it was tolerated by both the Chinese officials in Macau and the Portuguese. Then, over the years, in three stages Macau became the world's largest casino market. Stage 1 was the specific legalization of gambling in the 1840s for the purpose of raising government revenue. To meet its financial responsibility, in 1844, the Macau government required licenses for the various gambling establishments, the taxes from which were intended to help balance the budget. And so it was that taxes on gambling houses became a fiscal policy tool (Gaming, Inspection, and Coordination Bureau, Macao SAR, 2015).

Stage 2 in the evolution of gambling in Macau occurred in the 1930s, when the government of Macau changed from a licensing system, in which a large number of small gambling houses were franchised and paid tax, to a monopoly system, in which one syndicate controlled all and paid the entire tax. From time to time the monopoly was put up for bids. This gave the government better control over gambling and enabled it to obligate the monopoly syndicate to support the broader needs of Macau, for example, by promoting tourism. In 1962, the syndicate that won the monopoly then engaged in infrastructure improvements—dredging channels, providing high-speed ferry service from Hong Kong, and building hotels. At the same time, it introduced a number of European games in the casinos, giving them a Western look. By the time the Chinese regained sovereignty over Macau in 1999, gambling was well entrenched as a source of government revenue and a means by which broader community needs could be met (Gaming, Inspection and Coordination Bureau, Macao SAR, 2015).

Stage 3 in the evolution of legalized casino gambling came in 2002, when the relatively new Macau SAR government decided to liberalize and modernize the gaming industry there. The goal of liberalization was to bring competitive and state-of-the-art gaming practices to Macau for the purpose of developing the tourist industry, as well as giving a boost to the general economy at a time Macau was facing a slump in the construction industry and an unemployment problem (Ho, 2005, p. 43; Lam, 2002, p. 220). The plan was to move beyond just gaming-based tourism and to attract the nongaming tourist market and the so-called meetings, incentives, conventions and exhibitions (MICE) market. To serve these markets would not only require restructuring the gaming industry in Macau, it would also require improving Macau's image and reputation (McCartney, 2005, pp. 43–44).

Liberalization was launched by ending the existing monopoly and by inviting organizations with sufficient capital and extensive experience in the industry to bid on concessions. Three concessions were awarded in 2002. One went to a subsidiary of the former monopoly, and the other two went to US firms with major operations in Las Vegas. That two of the new concessions in Macau went to well-established Las Vegas firms is a clue to the Macau SAR strategy. Las Vegas was a model to copy. It serves a broad market that includes business and leisure travelers, and it serves the MICE market. Macau planned to have all of that (Gaming, Inspection and Coordination Bureau, Macao SAR, 2015). A top Las Vegas executive with one of the new concession holders said he was "hoping to transform Macau from a stop for just hard-core gamblers into a vacation destination" (Bashman & Kwon, 2008, p. 19).

Liberalization

The first new casino opened in 2004. By the end of 2011, there were 34 casinos in Macau where there had been only 11 in 2002. One more casino was in operation by 2015. As each of the three concessionaires was allowed to appoint another concessionaire, there are now a total of six concessions or franchises in Macau (Gaming, Inspection, and Coordination Bureau, Macao SAR, 2015).

The new casinos help attract tourists out of proportion to the local population. Over 28 million visitors arrived in Macau in 2011. Hong Kong, with more than 12 times the population, had fewer than 42 million tourists the same year. This increase in tourism can be attributed not only to gaming but also to other attractions in Macau, and to a revision in China's policy toward travel outside the mainland (Gaming, Inspection and Coordination Bureau, Macao SAR, 2015).

Macau tourism was further boosted in July 2003, when the government of China initiated the Individual Visit Scheme (IVS) that allowed mainland Chinese to travel to Macau as individual visitors rather than having to travel to Macau in a tourist group. In 2011, 40.8% of the visitors from mainland China were IVS visitors (Macau, 2011e).

For whatever reason the visitors have come to Macau, there has been a positive correlation between increases in tourism and increases in gross gambling revenue, and gross gambling revenue has increased at a faster rate than the increase in tourism. As shown in Table 1, the average annual rate of growth in tourism between 2003 and 2011, the year before the first new casino opened, has been 15.2%. The average annual rate of increase in gross gambling revenue for the same period was 31.4%. As a result of this rapid growth, Macau is now the largest gambling center in the world. By 2011, gross gambling revenue of Macau was 3.6 times as great as that of the Las Vegas Strip plus the rest of the county (Clark) in which the Strip is located.[6] Our calculations based on more recent data indicate this figure has now risen to 4.6 times the Las Vegas annual revenue. This has been prosperous for the casinos in Macau, and it has also been prosperous for the Macau government, as 81.0% of the government's revenue comes from taxes collected on gaming (Macau, 2012a).

Table 1. Economic development indicators: Macau and Hong Kong 1998–2011.

Economic indicator: average annual percent change	Years	Macau	Hong Kong
Real per capita GDP	1998–2002	1.2%	0.5%
	2003–2011	11.3%	4.2%
Employment	1998–2002	0.9%	0.4%
	2003–2011	5.4%	1.3%
Government revenue	1998–2003	3.4%	−2.6%
	2004–2011	26.8%	9.9%
Visitors	1998–2003	9.2%	5.5%
	2004–2011	15.2%	13.2%

Sources: (1) Macau (2011a, 2011b, 2011d, 2012a). (2) Hong Kong (2011, 2012, 1998–2012a, 1998–2012b).

During this period, there has been a major change in the image of Macau as it started to look more like Las Vegas. Yeung, Lee, and Kee (2008, p. 314) wrote:

> The most significant outcome was Macao's ability to attract experienced Las Vegas investors who introduced a successful integrated approach to gambling. Accordingly, the gaming operations began to combine opulence with hospitality, entertainment, convention facilities, and exhibitions suitable for visitors seeking excitement at the gambling tables as well as conventioneers, and tourists with families.

Toward that end, casino hotels such as the Venetian, Wynne, Galaxy, and Grand Lisboa have been built in Macau at great expense. Upscale shopping arcades can be found in all of them.

Despite Macau's modest size, the economic impact of the tremendous increase in gaming revenues has been substantial in the goal of making Macau a key destination for tourists and gamblers from China and Hong Kong as the data in Tables 2 and 3 clearly affirm. To what extent this growth has enabled Macau to be a pivotal node in the emerging PRD city-region as well as in a larger global network is questionable as we shall explain in the following narrative. We shall try to explore in greater depth the extent to which Macau has exerted any substantial thrust to the dynamism of the greater PRD's advance as a major Asian city-region and economic node in a larger global network.

VIP market

In the gaming industry, there is a distinction between the VIP market—for people who wager large sums of money, or so-called high rollers—and the mass market—for everyone

Table 2. Analysis of visitors by place of residence: Macau 2011.

	Macau visitors (000s)			
Place of residence	Total visitors	% total	Same-day visitors	% by residence
Mainland China	16,163	57.7	8,821	54.6
Hong Kong	7,583	27.1	4,165	54.9
Other Asia	3,541	12.6	1,739	49.1
All other	715	2.6	352	49.2
Total	28,002	100.0	15,077	53.8

Source: Macau (2011e).

Table 3. Change in number of visitors by place of residence: Macau 2008–2011.

	Macau visitors (000s)				
					% change
Place of residence	2008	2009	2010	2011	2008–2011
Mainland China	11,613	10,990	13,229	16,162	39.2
Hong Kong	7,016	6,728	7,466	7,583	8.1
Other Asia	3,532	3,343	3,578	3,542	0.3
All other	771	692	692	715	−7.3
Total	22,932	21,753	24,965	28,002	22.1

Source: Macau (2011f).

else. This difference is especially pronounced in Macau because of its unique VIP System.[7] The VIP System consists of casinos within casinos. The VIPs gamble in separate rooms managed by independent promoters under contract with the casinos. These rooms are marketed by independent junket operators who rely on personal contacts, or *guanxi*, to find customers.

Three salient features about the VIP customers in Macau are the following: (1) the system allows them a high degree of "anonymity, not possible in Western casinos" (Wang & Eadington, 2007, p. 33); (2) they are "gambling-only players who have little regard for other tourist amenities" (Wang & Eadington, 2007, p. 3); and (3) a large number of the VIPs, perhaps 50%, are "government officials or senior managers in state-run companies" who have access to bank credit on the mainland (CECC, 2011, p. 224). This is in spite of the fact that, according to Lo (2009), officials and state enterprise managers from China are forbidden to gamble in Macau. There are a number of questions about these VIP rooms owing to their lack of transparency. Capital control regulations limit the amount of money that a person can take out of China at 20,000 yuan per day (about US$3,262) (Hunt, 2013). This is a small amount for high rollers. The junket operators apparently front money to the gamblers who pay it back once they return to China. This may also be a method for laundering money to leave the People's Republic of China and Macau.

In most VIP rooms baccarat is the only game, and for the last 10 years gross gambling revenue from VIP baccarat alone has averaged 70% of the total gross revenue for all casinos. In 2011, the average was 73%. Because of the extra costs involved in VIP rooms, such as payments to promoters and junket operators, the net profit to the casinos is less than the net profit from the mass-market revenue. However, the government tax rate on all gross gambling revenue is the same, so the VIP system has serious implications for the government's budget.

Visitor profile

The visitors to Macau predominantly are from mainland China and Hong Kong (Gaming, Inspection and Coordination Bureau, Macao SAR, 2015). They also are predominately same-day visitors. The following two tables will show Macau's current and increasing dependency on regional short-term tourists. Table 2 presents data about the place of residence and number of visitors to Macau for the year 2011. The number and percentage of visitors from each place of residence is given. For example, in 2011, there were 16,162,700 visitors to Macau from mainland China, which were 57.7% of the total number of visitors. Of these visitors from mainland China, 54.6% were same-day visitors. From other information, when only the mainland Chinese visitors from Guangdong Province are counted,[8] Guangdong and Hong Kong together account for 56.4% of Macau's total visitors.

These data offer some important clues to the linkage between Macau and the PRD and may help explain further the success or lack thereof of Macau in helping to propel the PRD as a major city-region in the global economy. If slightly more than half of Macau's visitors are from Guangdong, the role of Macau is predominantly regional in scope. Macau may also be said to be important for its role in import substitution to prevent leakage of capital from Chinese nationals who go outside to gamble. At the same time, this would appear to suggest that the role Macau is playing as the recreational center for the coast of China is limited in its global appeal or in propelling the PRD as a city-region in the global economy.

Table 3 presents data on the total change in the number of visitors to Macau from 2008 to 2011, by place of residence. The increase in visitors from mainland China (39.2%) has been much greater than the increase of the total number of visitors from all other places. By the year 2014, the trend of increasing visitors from the mainland had continued, although with a slight decline in the first quarter of 2015. About 88% of Macau's visitors were from the mainland or Hong Kong, more than half of who were day-trippers.

Table 4 presents a comparison of visitor and gaming statistics for Macau and Las Vegas, indicating key differences between the developing gaming market in Macau and the long-established one in Las Vegas that Macau is attempting to emulate.

The data in Table 4 indicate the growth rate of visitors and gaming revenue in the two locations. The period 2003–2011 represents the period of rapid growth in Macau, while the period 2008–2011 represents the more recent trend. The growth in the number of visitors to Macau has slowed recently, but the rate of growth in gross gaming revenue

Table 4. Comparative visitor and gaming statistics: Macau-Las Vegas (Clark County) 2011.

	Macau	Las Vegas (Clark County)
Visitor arrivals 2011 (000)	28,002	38,929
Annual % increase		
2003–2011	15.2	1.4
2008–2011	6.9	1.3
Same-day visitors % of total	53.8	0.2
Overnight visitors % of total	46.2	99.8
Hotel rooms	22,356	150,161
Occupancy rate (%)	89.7	83.8
Average length of stay (nights)	1.5	3.7
Gaming revenue 2011 (US$mil)	33,632	9,223
Annual % increase		
2003–2011	31.4	2.1
2008–2011	34.8	−2.0
MICE participants 2011 (000)	1,278	4,865
MICE events	1,045	19,029

Sources: (1) Macau (2010–2012, 2011d, 2011f, 2012b). (2) LV (2011, 2012).

actually increased until the anti-corruption drive in China in 2014. The most remarkable decline has been in VIP baccarat gross revenue of 42.1% from the first quarter of 2014 to the first quarter of 2015. In Las Vegas, the growth rate in visitors has been steady, and gaming revenue has actually declined recently. The growth in visitors and gaming revenue has been substantially greater in Macau than in Las Vegas, suggesting that the Macau market is far from saturated. However, visitors to Las Vegas stay for a substantially longer period, and it requires relatively more hotel rooms in Las Vegas to accommodate them. Further, Macau is far behind Las Vegas in developing the MICE business. To further distinguish Macau from Las Vegas, the nongaming entertainment market in Macau seems to be weakening (Balsas, 2013; Simpson, 2009). According to Li (2012), an international and usually popular entertainment, Cirque du Soleil, located at a showcase casino hotel in Macau, had to close for lack of business. Li (2012) further reported "business owners blame lack of interest in anything apart from gambling for the inability to turn Macau into a destination market" (see also Simpson, 2009).

Economic vulnerability

The current gaming situation in Macau leaves it vulnerable on two counts. The first is lack of economic diversity because of the concentration in the gaming and related industries. Liberalization has brought about an increase in tertiary economic activity at the expense of secondary economic activity. Since 1999, there has been a precipitous decline in manufacturing jobs that once produced textiles, garments, toys, and fireworks, among other things. From 1999 to 2014, the number of jobs in manufacturing fell from 42,700 to 7,400. During the same period, the number of jobs classified as recreational, cultural, gaming, and other services increased from 19,300 to 94,000. In addition, the number of jobs in the gaming-related industry classification of hotel, restaurants, and similar activities increased from 21,000 to 54,800 (Macau, 2011c).

The second vulnerability is competition from other casinos in Asia or the possibility of another casino location opening in China (O'Keefe, 2014). Although the Macau casinos are probably the best known in East Asia, there is competition. There are casinos in Korea, the Philippines, and some Southeast Asian countries, although these tend to be small. More notable, two new casino resorts have opened in Singapore—one by a Las Vegas firm that also has a Macau concession (Holmes & Barta, 2012, p. B8). These two casinos have already had an impact on tourism in Singapore and could offer Macau strong competition, especially for the South and Southeast Asian markets as well as the Australia and New Zealand markets. As we shall discuss, changing policy with new leadership in China has created great challenges to Macau as the anti-corruption drive has greatly reduced the visits of Chinese VIP high rollers (O'Keefe, 2015).

The economic and social impact of casinos

The question of the impact of casino gambling, both economic and social, has been extensively studied without a consensus (Fong, Fong, & Li, 2011). There is a great deal of disagreement in the literature. These impact studies, supported by both statistical and anecdotal evidence, often come to opposite conclusions and are difficult because so many variables cannot be well defined and quantified. A report by the US National Gambling Impact Study Commission[9] summed it up. "The central issue is whether the net increases in income and well-being are worth the acknowledged social costs of gambling, ... it is currently impossible to obtain even a rough approximation of a true cost-benefit calculation of the economic impact of legalized gambling" (NGISC, 1999, pp. 7–29).

Casino types

Before examining the impact of casinos on society in general, and Macau in particular, it is helpful to classify them. The literature, in general, recognizes two basic types: the local casino and the tourist casino (Eadington, 1998, p. 60).[10] Not all casinos conform to these models, but the models are helpful in analyzing the economic and social impact of casinos.

The local casino is located in a metropolitan area. It uses local resources and serves the local community. Its customers are "day-trippers"—that is, they just come to the casino to gamble and then return home the same day. One possible benefit to the community is import substitution, or defensive legalization, in cases where members of the community gamble locally rather than visiting a tourist casino in another community (Walker, 2007, p. 36).

The tourist casino, sometimes referred to as a destination casino, is located away from populated areas or in a different jurisdiction from its customers. It uses local resources, but it exports gaming services to out-of-area customers.[11] Its business depends on tourism. Many of the customers will want overnight accommodations. This casino is typically a casino–hotel complex offering restaurants, retail stores, recreation, and nongambling entertainment. Because this casino is exporting its services, together with the multiplier effect, it can have a major positive impact on local income.[12]

Economic considerations

Casinos are attractive for many communities because of anticipated interrelated economic benefits, including (1) an increase in the general economic prosperity, (2) the creation of new jobs, (3) increased government revenue, and (4) the development of the tourist industry (Table 1). It is uncertain whether these economic benefits can be realized in every case. It depends largely on the specific circumstances for the specific casino and whether the casino is closer to the local model or the tourist model.

Casinos have the potential to raise the average standard of living in a community, for which real per-capita gross domestic product (GDP) can serve as a proxy. The increase will be more pronounced the closer the casinos are to the tourist model and will be the result of tourist spending, which in turn will create local jobs. In many cases, it is the potential tax revenue from gaming that motivates political jurisdictions to legalize gaming (Walker, 2007, p. 9). Taxes and other fees the government receives from the casinos will tend to lower taxes for others and enable the government to provide more social goods than otherwise. However, if the local government becomes too dependent on tax revenue from casinos, it could have a severe financial shock if there is a downturn in gaming. In 2014, approximately 80% of Macau's tax revenue came from taxes on gaming (Gaming, Inspection and Coordination Bureau, Macao SAR, 2015).

Social considerations

Social considerations are more difficult to define and quantify than the purely economic considerations (Fong et al., 2011). Casino gambling is said to provide a social benefit for people who gamble in moderation and enjoy the experience (Eadington, 1999b, p. 186).

On the other hand, casino gambling is associated with a variety of social costs. The major social costs can be classified as criminal activity, and problem and pathological gamblers (Eadington, 1999a, p. 134).

One aspect of criminal activity is violence. In 1999, immediately before Macau was returned to Chinese sovereignty, there was a major outbreak of violence and 37 people died in gang-related killings in Macau. The violence subsided after the handover, but it has recently flared up again. In July 2012, three mainland Chinese were killed in or near casino-hotel resorts in Macau (Wall Street Journal, 2012). Additional assaults in the area resulted in a coordinated Macau–Hong Kong–mainland China raid on hotels and casinos in which 1,300 people were questioned and 150 arrested (Bradsher & Gough, 2012; New York Times, 2012).

Another aspect of criminal activity is financial crimes. There is a large amount of money associated with gambling, and the nature of the business creates opportunities for various financial crimes including skimming, illegal betting, and money laundering. These are not activities new to Macau, but they have increased in intensity with the expansion of the gaming industry. "The growth of gambling in Macau … has been accompanied by widespread corruption, organized crime, and money-laundering" (CECC (Congressional-Executive Commission on China), 2011, p. 51; see also Hing, 2005). The VIP rooms are especially attractive to organized crime because of their veil of secrecy and the large amounts of money wagered there. Further, money laundering in Macau has been an issue that goes beyond just casinos and VIP patrons. In 2006, the United States intervened in a North Korean scheme that was laundering money through Macau banks.

A third aspect of criminal activity is human trafficking. There is a history of such abuse in Macau at least going back to the coolie trade of the nineteenth century. There is an equally long history of human trafficking for sex that has persisted until today. In Macau, prostitution is not a crime, but sex trafficking is, and the touring and gaming industries make Macau an attractive destination for sex trafficking (Macau SAR Government, 2011, p. 45). The influx of mainland visitors "has been accompanied by a visible influx of mainland prostitutes" (Fraser & Lo, 2012; Lo, 2009, p. 139; USDS, 2012). This trafficking is often done by organized crime gangs.

These crimes often involve a cross-border network so that all of the social costs of gambling may not occur in the same place that the economic benefits occur. One of the findings of the National Gambling Impact Study Commission (NGISC, 1999, pp. 7–11) was that "the economic benefits are generally more pronounced within the immediate vicinity of the gambling facility, while the social costs tend to be diffused throughout a broader geographical region."

Problem and pathological gambling

Apart from the criminal activity described above are crimes and other negative consequences associated with problem and pathological gambling, which is recognized as an impulse control psychiatric disorder.[13] The individuals afflicted are addicted to gambling. Problem and pathological gamblers are associated with emotional instability, antisocial and disruptive behavior, dysfunctional families and divorce, health issues including mental health, alcohol and drug abuse, suicide, financial difficulties, employment difficulties, and problems with the law (Volberg, 1994).

Problem and pathological gambling are latent problems that emerge when gambling is at hand. Volberg (1994, p. 239) concluded "increasing the availability of gambling will contribute to an increase in the prevalence of gambling-related problems in the general population." Fong (2012) reported that between 2003 and 2007, the prevalence rate of problem and pathological gambling in Macau increased from 2.5% and 1.8% to 3.4% and 2.6%, respectively. Wan, Li, and Kong (2001) reported that between the end of 2004 and the end of 2008, the increase in gambling-related crime was greater than the increase in total crime. Further, they also reported that after a Macau hotline counseling service was made available in 2005, almost 90% of the calls received were for gambling-related problems.

The disruption caused by problem and pathological gamblers is not limited to the gamblers but "spills over into the household, other family members, friends, employers, creditors, and the community as a whole" (Gerstein et al., 1999, p. 38). These gamblers create "harmful by-products of gambling that affect both those who gamble and those who do not ..." (Grinols, 2004, p. 131). This places an additional burden on public health facilities and law enforcement agencies and, as a result, places an economic burden on society even if the burden cannot be precisely measured.

Analysis and interpretation

Macau and its place in the PRD is both exceptional and unique, for there is no city in the world that has its special and extraordinary character as gambling center to the world's largest population and recreational/tourist anchor to a mega-urban region of more than 40 million people (Simpson, 2008). In analyzing and interpreting its continuing evolution, it is appropriate to go back to the three research questions or goals we asked at the beginning of the paper. In this way, we can frame Macau's ongoing transaction as urban gambling broker and recreational center for southeast China as well as the PRD region.

These research questions focused on the economic and social impacts of liberalizing gaming (gambling) on tourism and how these have advanced the goals of making Macau the key recreation/tourist center of the PRD urban region. Related to this is a broader question linked to the conceptual idea of city-regions and the success or failure of Macau to galvanize a key economic driver of tourism as part of the emerging global trajectory of the PRD city-region. We specifically sought to look not just at the positive benefits but also to see if there are significant negative consequences of the remarkable growth in gambling activities and revenues. Finally, we asked if Macau has met the expectations of China's policy makers and planners as articulated in the 2009 comprehensive plan for its role in the PRD city-region.

The economic impacts have been substantial and far-reaching. Enormous new streams of revenue have become available, and the results can be seen in the upgraded infrastructure, increases in government revenue, and the much higher per-capita incomes from the years 2003–2011, as noted in Table 1. By any reasonable measure, the economic impact has been striking. Job growth in the casinos and hotels has benefited the local population, and residents appear to be much better off on virtually all economic indicators.

The social impacts are not as clear-cut, but such impacts appear to be largely negative. Criminal activity continues, and violent crimes have not been eliminated as killings and beatings occur. Prostitution is rampant in the casinos and hotels and will only grow as the gaming activity grows and the numbers of casinos increase. With so much money involved through the huge increases in gambling revenues, various crimes associated with money and finance especially related to the gambling in the VIP rooms of the casinos have become a very serious problem (Lo, 2009). Triads based in both Macau and the mainland vie for participation in this very lucrative share of the gaming take, and collecting debts across the SAR/China border leads to sometime violent exchanges.

Despite any gambling restrictions that may apply to mainland Chinese officials and state enterprise managers, many arrive nominally on business trips or perhaps coming from Hong Kong for a day of sightseeing. Where do they obtain the large sums of money to gamble in the VIP rooms and why do they engage in such activity? Money can be borrowed and provided by the junket operators and what Lo (2009, p. 140) describes as the "chip rollers" in Macau who make their money by facilitating and servicing the high-roller gamblers from China. Today, these high rollers and Chinese officials and entrepreneurs face the real danger of being investigated for corruption in the wake of the new policies of President Xi Jinping. The era of VIP rooms and the secrecy surrounding them may be over, although it is too early to know the outcome.

In seeking to answer the third research question, we argue that Macau in fact has not created the kind of "Gold Coast to become the most attractive tourism and leisure center … in the world" as envisioned in the 2009 plan. The reality is that all the data we have presented in Tables 2 and 4 indicate that Macau is mainly a location for short-term gamblers from the China mainland and Hong Kong, many of who are day-trippers. However, Macau does not offer as a vacation destination for families and conventioneers, nor does it offer the variety and scope of entertainment and services found in Las Vegas. Consequently, we argue that Macau has not contributed significantly to the trajectory of the PRD as an emerging super city-region in the global economic system with extensive connections to regions outside East and Southeast Asia. In answer to Simpson's (2008) somewhat rhetorical question of Macau, as capital of the twenty-first century, our findings suggest that it has not achieved that success or likely outcome.

Is Macau living up to the expectations of what Chinese policy makers and planners envisioned when they decided to allow the modernization and expansion of gaming in Macau? If they were serious in their expectation that it would become the greatest vacation and tourist destination in the world, the answer is a resounding "no." The recent anti-corruption actions of President Xi Jinping in visiting Macau in December 2014 and his new directives to Macau to clean up and to develop tourist and nongaming activities on neighboring Hengqin Island demonstrate clearly that Macau did not live up to the expectations of China's leader and the central state (O'Keefe, 2015; Shih, Fraser, & Carvalho, 2015).

On the other hand, if there was in fact a less obvious agenda in encouraging Macau to become the world's largest and perhaps most impressive gaming center, accessible to millions of Chinese right on China's doorstep and a place where China can control the gambling as well as ensure the import substitution aspect of controlling the capital flows, then perhaps Macau did become what was expected.

In addition, the economic success of Macau in creating such huge and continuing growth in gaming revenues and tax receipts has resulted in remarkable new prosperity for a subregion of the PRD unlike anything Macau and its adjacent city of Zhuhai have ever witnessed. While not being a salient ingredient in the success of the PRD, it has certainly contributed meaningfully to its prosperity and economic growth. Yet the future of such growth remains uncertain in the wake of the current anti-corruption drive and policy in today's China.

Summary and conclusion

Macau remains one of the most intriguing cities in the world and something of an anomaly among Chinese cities. On the one hand, the heritage of a former Portuguese territory and colony is diminishing, but the culture and history of Macau's prolonged search for financial resources has resulted in continuing the tradition of gambling, its most enduring source of capital. Never has Macau witnessed such prosperity as its explosion of gambling casinos and world-class hotels has brought.

Yet Macau is primarily evolving into a local casino city serving the huge PRD city-region. The visitors, as noted, are primarily same-day visitors. In 2011, the number was 53.8% of the total and rose to 56.6% in the first quarter of 2015, and this does not include local Macau citizens. When compared to Las Vegas, it has relatively few hotel rooms, and its convention trade has not matured. It has a weak entertainment market, and most of its visitors are not attracted to things outside of gambling. Its gambling market is two-tiered —VIP and mass market—and the main gaming revenue is derived from the VIP market. This VIP market is now under severe threat from new policies emanating from China's President as he seeks to weed out corruption among officials. It is difficult to know if this will be a permanent blow to the growth of gaming revenue in Macau or if this new policy will dissipate after a time and gradually return to the high growth days.

Macau's role as a major gaming center is indeed a curious thing for the People's Republic of China to sponsor and encourage. While it has been done under state supervision, it has required a minimum of state investment owing to the liberalization of gaming that allowed foreign money to invest in these enterprises. As Naughton (2011) has noted, China, following the global financial crisis of 2008, has modified its focus on market-based incentives to drive its economic growth and has to some extent reverted to a more state-oriented and state-managed control of its economy, especially in such key sectors as energy, transportation, health care, and finance.

The idea of state intervention has taken on a much larger role in recent years and this has led some to ask if China does indeed have its own hybrid model for economic growth that diverges from the conventional Western model (Lin & Hu, 2011). Yet its SARs (Macau and Hong Kong) have been allowed or indeed encouraged to proceed on their own, with the assumption that they may keep their own management and good standard of living if they can pay their own way. At the same time, they are nudged to proceed with figuring out the best path to future integration back into the mainland and their local PRD region.

While this is a still evolving situation, it can be seen as a fascinating case study in the conceptual context of McCann and Ward's (2011) ideas of mobile urbanism in which there is constant interplay between the local (Macau) and the national (China's central state) as well as the global (the goal of making Macau a world center of gaming, recreation, and conferencing). The powerful effect of new policies emanating from Beijing and the manner in which Macau and its chief executive and government respond and the effects on the ground in Macau are now unfolding. The recent impacts this has on the declining profits of multinational gaming companies such as Sands, MGM, and Wynn are clear, and there will no doubt be a paralleling impact on the people and infrastructure of Macau, a fascinating validation of the reality of global neoliberalism and its confrontation with a strong central state in China.

In the context of such thinking about China's current economic trajectory of slowing growth, where do we position the nature of its urban trajectory and the place of a city such as Macau? The inexorable forces of China's massive urban transition have swept close by as Macau's neighbor Zhuhai has far surpassed its population and overshadowed it with its much larger territory and urban expansion. Yet Macau's distinctive history and now state-directed role as China's key gaming center and tourist destination in the PRD city-region likely ensure its future viability and exceptional status as the world's greatest urban casino but one with an occasional volatile eruption.

Disclosure statement

No potential conflict of interest was reported by the authors.

Notes

1. In this paper, the spelling "Macau" will be used except in instances where the source document uses "Macao."
2. The 11 principal cities are Guangzhou, Shenzhen, Zhuhai, Foshan, Dongguan, Zhongshan, Jiangmen, Huizhou Zhaoqing, Macau, and Hong Kong.
3. The Greater PRD includes Macau SAR and Hong Kong SAR.
4. Other specific objectives include the creation of a Macau–Zhuhai metropolitan area, the development of Hengqin Island adjacent to Macau, together with Zhuhai the creation of a Guangzhou–Zhuhai–Macau axis to help develop the western side of the PRD, and the development of an economic liaison with Portuguese-speaking countries (Guangdong Province, Hong Kong SAR, & Macau SAR, 2009, p. 40).
5. More details on the history of gambling in Macau may be reviewed in Loughlin and Pannell (2010a). See also Gaming, Inspection and Coordination Bureau, Macao SAR (2015).
6. LVA reported a total of US$9,223 million of gross gaming revenue for Clark County, Nevada, in 2011. Macau (2012a) reported gross gaming revenue (games of chance) for US$33,632 million in 2011.
7. See Wang and Eadington (2007) for the details of the Macau VIP System.
8. There were 8,196,139 visitors to Macau from Guangdong Province in 2011. See Macau (2011e).
9. The National Gambling Impact Study Commission was established by the US Congress in 1996 to conduct a study of gambling in the United States.
10. In much of the literature, the local casino, also known as an urban casino or a casino of convenience, is made analogous to a restaurant that uses local resources and serves a local market. The tourist casino, sometimes referred to as a destination casino, is made analogous to a factory that uses local resources but exports its products out of the local market.
11. Services provided to tourists are considered an export to the tourists' home.
12. A multiplier magnifies the impact of an injection of income into an economy from outside (such as from tourists) by creating a chain reaction of income and spending in the local economy.
13. See DSM-IV-TR (2000). The difference between problem and pathological gamblers is a difference in degree, measurable by psychiatric tests.

References

Associated Press. (2014, December 20). In wake of Hong Kong, Chinese leader warns Macau. *International New York Times*. Retrieved from http://nyti.ms/1DWrZad

Balsas, Carlos J. L. (2013). Gaming anyone? A comparative study of recent urban development trends in Las Vegas and Macau. *Cities*, *31*, 298–307.

Bashman, Mark, & Kwon, Esther Y. (2008). Industry surveys: Lodging & gaming. *Standard & Poor's industry surveys*. New York, NY: McGraw-Hill.

Bradsher, Keith, & Gough, Neil. (2012, June 28). Attack raises fears of a new gang war in Macau: Investor is beaten at his own casino. *The New York Times*. Retrieved from http://www.query.nytimes.com

CECC (Congressional-Executive Commission on China). (2011). *Annual report 2011*. Washington, DC: US Government Printing Office. Retrieved November 6, 2012, from http://www.cecc.gov/pages/annualRpt/annualRpt11/AR2011final.pdf

Chan, Kam Wing. (2012). Crossing the 50 percent population rubicon: Can China urbanize to prosperity? *Eurasian Geography and Economics*, *53*(1), 63–86.

Chow, Gregory C. (2002). *China's economic transformation*. Malden, MA: Blackwell Publishing.

Dicken, Peter. (2007). *Global shift: Mapping the changing contours of the world economy* (5th ed.). New York, NY: The Guilford Press.

DSM-IV-TR (Diagnostic and statistical manual of mental disorders). (2000). *Text revision* (4th ed.). Washington, DC: American Psychiatric Association.

Eadington, William R. (1998). Contributions of casino-style gambling to local economies. *The Annals of the American Academy of Political and Social Science*, *556*, 53–65.

Eadington, William R. (1999a). The economics of casino gambling. *Journal of Economic Perspectives*, *13*(3), 173–192.

Eadington, William R. (1999b). The spread of casinos and their role in tourism development. In Douglas G. Pearce & Richard W. Butler (Eds.), *Contemporary issues in tourism development* (pp. 128–143). New York, NY: Routledge.

Enright, Michael J., Scott, Edith E., & Chang, Ka-mun. (2005). *Regional powerhouse: The greater Pearl River Delta and the rise of China*. Singapore: John Wiley & Sons (Asia).

Fong, Davis. (2012). Responsible gambling in Chinese community—experience from Macau. A presentation at the Asian Casino & Gambling Congress 2010 held at Marian Bay Sands. Singapore, October 5–7.

Fong, Davis K. C., Fong, Hoc Nang, & Li, Shao Zhi. (2011). The social cost of gambling in Macao: Before and after the liberalisation of the gaming industry. *International Gambling Studies*, *11* (1), 43–56.

Fraser, Niall, & Lo, C. (2012, December 10). Vice squad raids casino hours after owner Adelson flies in. *The South China Morning Post*, p. 10. Retrieved from http://www.scmp.com

Gaming, Inspection and Coordination Bureau, Macao SAR. (2015). *Macao gaming history*. Direcção de Inspecção e Coordenação de Jogos (DICJ). Retrieved from http://www.dicj/gpv.mo/web/en/history/index/html

Gerstein, Dean, et al. (1999). *Gambling impact and behavior study: Report to the National Gambling Impact Study Commission*. National Opinion Research Center at the University of Chicago, Gemini Research, and The Lewin Group. Retrieved September 16, 2010, from http://www.norc.org/new/gamb-fin.html

Ginsburg, Norton, Koppel, Bruce, & McGee, Terry G. (Eds). (1991). *The extended metropolis: Settlement transition in Asia*. Honolulu, HI: University of Hawaii Press.

Gottmann, Jean. (1961). *Megalopolis: The urbanized northeast seaboard of the United States*. Norwood, MA: Plimpton Press.

Grinols, Earl L. (2004). *Gambling in America: Costs and benefits*. Cambridge: Cambridge University Press.

Guangdong Province, Hong Kong SAR, & Macau SAR. (2009). *Building coordinated and sustainable world-class city-region. Planning study on the co-ordinated development of the greater Pearl River Delta townships*. Consolidated final report. Hong Kong: Author. Retrieved June 5, 2010, from http://www.pland.hk/pland_en/misc/great_prd/news/report-e.htm

Hing, Lo Shiu. (2005). Casino politics, organized crime and the post-colonial state in Macau. *Journal of Contemporary China*, *14*(43), 207–224.

Ho, W. S. (2005). Macao's economy since the handover. *AMCM Quarterly Bulletin*, January, 25–49. Retrieved November 10, 2012, from http://www.amcm.gov.mo/publication/quarterly/Jan2005/5yearReview_en.pdf

Holmes, Sam, & Barta, Patrick. (2012, October 5). Singapore casino experiment pays off. *Wall Street Journal*. Retrieved November 10, 2012, from http://online.wsj.com/article/SB10001424052748704680504575533793000481362.html

Hong Kong. (1998–2012a). *Annual digest of statistics*, various issues. Government of the Hong Kong Special Administrative Region, Census and Statistics Department. Retrieved from http://www.censtatd.gov.hk

Hong Kong. (1998–2012b). *Monthly digest of statistics*, various issues. Government of the Hong Kong Special Administrative Region, Census and Statistics Department. Retrieved from http://www.censtatd.gov.hk

Hong Kong. (2011). *2011 gross domestic product*. Government of the Hong Kong Special Administrative Region, Census and Statistics Department. Retrieved from http://www.censtatd.gov.hk

Hong Kong. (2012). *Hong Kong in figures 2012*. Government of the Hong Kong Special Administrative Region, Census and Statistics Department. Retrieved from http://www.censtatd.gov.hk

Hsu, Ann Y., Pannell, Clifton W., & Wheeler, James O. (1980). The development and structure of transportation networks in Taiwan: 1600–1972. In Ronald G. Knapp (Ed.), *China's island frontier: Studies in the historical geography of Taiwan* (pp. 167–201). Honolulu, HI: The University Press of Hawaii.

Hunt, Katie. (2013). The dark side of Asia's gambling Mecca. *CNN.com*. Retrieved June 18, 2013 from http://www.cnn.com/2013/06/17/world/asia/macau-dark-side/index.ht

Lam, Newman M. K. (2002). Government intervention in Macao's economy. *Asian Journal of Public Administration*, *24*(2), 211–233.

Li, Feng, & Li, Jing. (1999). *Foreign investment in China*. London: MacMillan Press.

Li, Z. (2012). Show's over: Cirque du Soleil tumbles out of Macau. *CNN International*. Retrieved November 12, 2012, from http://www.cnngo.com/hong-kong/visit/cirque-du-soleil-macau-871570

Lin, George C. S. (1997). *Red capitalism in south China: Growth and development in the Pearl River Delta*. Vancouver, BC: UBC Press.

Lin, George C. S., & Hu, Fox Z. Y. (2011). Getting the China story right: Insights from national economic censuses. *Eurasian Geography and Economics*, *52*(5), 712–746.

Lo, Shiu S. (2009). *The politics of cross-border crime in greater China: Case studies of Mainland China, Hong Kong, and Macao*. Armonk, NY: M.E. Sharpe.

Loughlin, Philip, & Pannell, Clifton W. (2010a). Gambling in Macau: A brief history and glance at today's modern casinos. *Focus on Geography*, *53*(1), 1–9.

Loughlin, Philip, & Pannell, Clifton W. (2010b). The Port of Hong Kong: Past successes, new realities and emerging challenges. *Focus on Geography*, *53*(2), 50–58.

LV. (2011). *2011 Las Vegas visitor profile*. Las Vegas Convention and Visitor Authority. Retrieved from http://www.lvcva.com

LV. (2012). *Historical Las Vegas visitor statistics (1970–2011)*. Las Vegas Convention and Visitor Authority. Retrieved from http://www.lvcva.com

Ma, Jun. (2000). *The Chinese economy in the 1990s*. New York, NY: St. Martin's Press.

Macau. (2010–2012). *Quarterly gaming statistics*, various issues. Macau SAR, Gaming Inspection and Coordination Bureau. Retrieved from http://www.dicj.gov.mo

Macau. (2011a). *Quarterly gross domestic product*, various issues. Government of Macao Special Administrative Region, Statistics and Census Service. Retrieved from http://www.dsec.gov.mo

Macau. (2011b). *Employed population by gender and age group*. Government of Macao Special Administrative Region, Statistics and Census Service. Retrieved from http://www.dsec.gov.mo

Macau. (2011c). *Employed population by industry*. Government of Macao Special Administrative Region, Statistics and Census Service. Retrieved from http://www.dsec.gov.mo

Macau. (2011d). *Principal statistical indicators of Macau*, various issues. Government of Macao Special Administrative Region, Statistics and Census Service. Retrieved from http://www.dsec.gov.mo

Macau. (2011e). *Tourism statistics*, various issues. Government of Macao Special Administrative Region, Statistics and Census Service. Retrieved from http://www.dsec.gov.mo

Macau. (2011f). *Tourism indicators*, various issues. Government of Macao Special Administrative Region, Statistics and Census Service. Retrieved from http://www.dsec.gov.mo

Macau. (2012a). *Monthly bulletin of statistics*, September. Government of Macao Special Administrative Region, Statistics and Census Service. http://www.dsec.gov.mo

Macau. (2012b). *MICE statistics 2011, 4th quarter*. Government of Macao Special Administrative Region, Statistics and Census Service. http://www.dsec.gov.mo

Macau SAR Government. (2011, October 17). *Report on the study of human trafficking in Macao*. Human Trafficking Deterrent Measures Concern Committee, p. 45. Retrieved October 16, from http://anti-tip.gov.mo/en/ news_detail.aspx?a_id=77

McCann, Eugene, & Ward, Kevin. (2011). Introduction, urban assemblages. In Eugene McCann & Kevin Ward (Eds.), *Mobile urbanism, cities and policy making in the global age* (pp. xiiv–xxxv). Minneapolis: University of Minnesota Press.

McCartney, Glenn J. (2005). Casinos as a tourist development strategy—The case of Macao. *Journal of Macau Gaming Research Association*, *2*, 40–54.

McGee, Terry G. (1991). The emergence of *Desakota* regions in Asia: Expanding a hypothesis. In Norton Ginsburg, Bruce Koppel, & Terry G. McGee (Eds.), *The extended metropolis: Settlement transition in Asia* (pp. 3–26). Honolulu, HI: University of Hawaii Press.

Murphy, Alexander B. (2012). China's cultural and creative economy: An introduction. *Eurasian Geography and Economics*, *53*(2), 179–181.

Naughton, Barry. (2011). China's economic policy today: The new state activism. *Eurasian Geography and Economics*, *52*(3), 313–329.

New York Times. (2012, August 7). World briefing. Macau: 150 arrested at casinos. *New York Times*, p. 6A.

NGISC (National Gambling Impact Study Commission). (1999). *National Gambling Impact Study Commission final report*. Washington, DC. Retrieved November 6, 2012, from http://govinfo.library.unt.edu/ngisc/reports/fullrpt.html

O'Keefe, Kate. (2014, December 17). Chinese president to visit Macau, where stakes are high. *Wall Street Journal*. Retrieved from http://www.wsj.com/articles.chinese-president-to-visit-macau-where-stakes-are-high-1418850002

O'Keefe, Kate. (2015, May 5). Macau's casino projects feel a chill. *Wall Street Journal*. Retrieved from http://www.wsj.com/page2_0333-20150505.html.

Office of the Chief Executive of the MSAR Government. (2015). Policy address for the fiscal year 2015 of the Macao Special Administrative Region (MSAR) of the People's Republic of China [translated version]. Retrieved from http://www.gce.goc.mo

Oshima, Harry. (1987). *Economic growth in monsoon Asia: A comparative survey*. Tokyo: Tokyo University Press.

Pannell, Clifton W. (2012). Hong Kong and Macau, twin dynamos in the Pearl River Delta. *China Currents*, *10*, 85–93.

Scott, Allen J. (2001). Global city-regions. In A. J. Scott, et al. (Eds.), *Global city-regions: Trends, theory, and policy* (pp. 11–30). Oxford: University of Oxford Press.

Shih, Tan Hock, Fraser, Niall & Carvalho, Raquel. (2015, January 18). How the communist party knocked down Macau's house of cards in just six weeks. *South China Morning Post*. Retrieved from http://www.scmp.com/print/news/china/article/1681571/party-knock

Simpson, Timothy. (2008). Macao, capital of the 21st century? *Environment and Planning D: Society and Space*, *26*(6), 1053–1079.

Simpson, Timothy. (2009). Materialist pedagogy: The function of themed environments in post-socialist consumption in Macao. *Tourist Studies*, *9*(1), 60–80.

US Government. (2015). *CIA handbook of countries. The world factbook*. Washington, DC: Author.

USDS (United States Department of State). (2012, June 19). *2012 Trafficking in Persons Report—Macau*, p. 228. Retrieved October 17, 2012, from http://www.unhcr.org/refworld/docid/4fe30cb23c.html

Veeck, Gregory, Pannell, Clifton W., Smith, Chris J., & Huang, Youqin. (2011). *China's geography, globalization and the dynamics of political, economic, and social change*. Lanham, MD: Rowman and Littlefield.

Vogel, Ezra F. (2011). *Deng Xiaoping and the transformation of China*. Cambridge, MA: Belknap Press.

Volberg, Rachel. (1994). The prevalence and demographics of pathological gamblers: Implications for public health. *American Journal of Public Health*, *84*(2), 237–241.

Walker, Douglas M. (2007). *The economics of casino gambling*. Berlin: Springer.

Wall Street Journal. (2012, August 6). Macau police arrest 150 in casino raids. *Wall Street Journal*. Retrieved November 4, 2012, from http://online.wsj.com/article/SB10000872396390443659204577572650198182324.html

Wan, Penny Y. King, Li, Crystal Xin, & Kong, Weng Hang. (2001). Social impacts of Casino gaming in Macao: A qualitative analysis. *Tourism: An International Interdisciplinary Journal*, *59*(1), 63–81.

Wang, Wuyi, & Eadington, William E. (2007, January). *VIP-room contractual system and Macau's traditional casino industry*. Economics Working Paper Series (Working Paper No. 07-001). Reno, NV: University of Nevada, Reno.

Yang, Chun. (2006). The geopolitics of cross-boundary governance in the Greater Pearl River Delta, China: A case study of the proposed Hong Kong-Zhuhai-Macao Bridge. *Political Geography*, *25*, 817–835.

Yeung, Yue-man, & Hu, Xuwei. (1992). *China's coastal cities: Catalysts for modernization*. Honolulu, HI: University of Hawaii Press.

Yeung, Yue-man, Lee, Joanna, & Kee, Gordon. (2008). Hong Kong and Macao under Chinese sovereignty. *Eurasian Geography and Economics*, *49*(3), 304–325.

Planning Beijing: socialist city, transitional city, and global city

Chaolin Gu[a], Yehua Dennis Wei[b,c] and Ian G. Cook[d]

[a]*Department of Urban Planning, Tsinghua University, Beijing, China;* [b]*Department of Land Management, Zhejiang University, Hangzhou, China;* [c]*Department of Geography, University of Utah, Salt Lake City, UT, USA;* [d]*School of Humanities and Social Science, Liverpool John Moores University, Liverpool, UK*

Cities are centers of economic, social, and political change, and urban planning is a process responding to and guiding urban change and development. In the Maoist era and under the influence of socialist ideology, China limited urbanization while promoting industrialization, and urban planning served as an instrument for socialist construction. Since the reform of the late 1970s, Chinese cities have experienced unprecedented growth and restructuring. However, the gradualist, exploratory reform —exemplified by Deng Xiaoping's slogan "crossing the river by feeling the stones"— makes Chinese cities constantly change without clear directions for future development. This paper uses Beijing as a case study to analyze changing institutional and global contexts underlying the transformation of Chinese cities, and planners' responses and dilemmas in making plans and implementing them. We found that market reforms, rapid growth, and dramatic change make urban master plans quickly out of date, forcing Chinese planners to frequently revise these master plans. We also found that the content of urban master planning in China has broadened from physical planning, and Chinese planning has adapted to market reform through utilizing concepts of visioning, flexibility, and governance. Increasingly what we call a "hybrid" form of planning is arising in which global concepts and Chinese ideas interweave in order to direct the shape and form of the Chinese metropolis.

1. Introduction

Rapid urban growth and spatial restructuring in developing countries, especially Asia, have generated considerable governmental concerns and scholarly attention (Hamnett & Forbes, 2011; Kasarda & Parnell, 1993; Park & Wheeler, 1983; United Nations, 1996, 2010). Urban development issues are important to governments as they are associated with uneven development, social problems, intergovernmental relations, and sustainability. Former socialist countries, under the influence of socialist ideology, intended to control urbanization and the growth of large cities, as evidenced by the notion of "under-urbanization" and "de-urbanization" policies (Ma, 2002; Wei, 1994). On the other hand, cities are centers of economic growth, political power, and international connections. In the context of globalization and competitiveness, urban governments are being transformed from providers of public goods to developmental and entrepreneurial states that actively promote urban growth and competitiveness (Hall & Hubbard, 1998;

Wu, 2002). In former socialist countries, economic reforms have empowered growth-oriented local governments, and cities in transitional economies, or transitional cities have become centers of urban growth and change. Market-oriented reforms have also transformed the ideology and practice of planning in former socialist countries.

During Mao's period after the founding of socialist China in 1949 and influenced by socialist ideology, China implemented a policy of controlling migration and urbanization, while promoting industrialization (Ma, 2002; Zhou & Ma, 2000). Chinese cities became centers of production rather than consumption, and in the late 1970s and early 1980s, China officially announced an urbanization policy to control the size of large cities (with non-agricultural population over 500,000). However, with reforms and globalization, Chinese cities are undergoing dramatic growth and restructuring. Chinese cities are being transformed from socialist productive cities to transitional cities and globalizing cities, and leading metropolises such as Shanghai and Beijing are seeking to become global cities (Cook, 2006; Walcott & Pannell, 2006; Wei & Yu, 2006). Urbanization and urban land expansion are unprecedented in Chinese history, and millions of rural migrants have moved to cities, up to a current total of 230 million according to some estimates. This rapid urbanization and industrialization imposes serious challenges of livability and sustainability in Chinese cities. In order to tackle this rapid growth, urban planning has become revitalized. However, in the face of this rapid pace of transformation, scholars are questioning the nature and effectiveness of urban planning in China (Luo & Shen, 2008; Wei, 2005; Wu, 2010).

This article analyzes changing urban planning in Chinese cities through a study of shifting planning contexts, planning revisions, and the development reality in Beijing. Beijing serves as a good case study for the following reasons. First, Beijing is known for its rise in the global city network and represents the challenges confronting cities in China in the midst of integration into a global production system (Wei & Yu, 2006). Second, Beijing demonstrates the general challenge of urban planning in China's transitional cities where urban growth is supposed to be under much stricter control of the central and local governments, not just the centrally administered cities of Beijing itself, Shanghai, Tianjin, and Chongqing, but also other cities such as Guangzhou or Hangzhou that are administered via provincial governments (Xie, Fang, Lin, Gong, & Qiao, 2007; Zhao, 2011). Third, although scholars have written about urban planning in Beijing (e.g., Gu & Cook, 2011; Sit, 1995, 1996; Tang, 2006; Zhao, 2011), the broader social-economic contexts that influence such changes have not been fully examined. Attention has also been paid to the rationale of plan making, while the differences between planning and development reality have not been detailed.

In the following sections, we will first overview institutional change and evolution of the urban planning system and approaches in China. This is followed by an investigation of the context and content of urban planning in Beijing to analyze the changing nature of urban planning. We will then analyze the mismatches between urban master plans and development realities. We argue that constant planning revision is rooted in the exploratory, gradualist nature of economic reforms, and that planning Chinese cities is an extremely difficult practice still in the process of reform and globalization. Although we are analyzing the contemporary era, it is important to appreciate that there is a legacy from the historical past, especially via "governmental control of society", as Abramson (2007, p. 68) notes.

2. China: institutional change and urban planning

Urban planning in China had a stuttering start in the early twentieth century, with what we term a "protomodern" introduction of some planning concepts from Western society in the early decades of that century. Newly built areas outside the ancient city were influenced by Western concepts, while the strong nationalist urge of the period presaged recent changes via the search for a synthesis between Chinese and Western cultures (Zhu, 2009). Following the founding of the People's Republic of China in 1949, planning became a key element of urban development and for the first 30 years under socialism followed the former Soviet Union's urban planning model based on the socialist planned economic system in which urban planning was a component of state economic development plans (Gu & Cook, 2011). Urban planning in China was associated with the utopian view of a classless society and focused on planned industrial growth in order to improve the income level of the masses (Wu, 2007). Urban planning was used to guarantee the socioeconomic plan and to materialize national economic planning in order to strengthen the development of the Chinese state (Tang, 2000). The content of master plans included (1) allocation of medium and large industrial construction projects, often near the inner city workforce; (2) urban functional areas; and (3) urban transportation and infrastructure. It included an overall spatial layout and some types of specialized planning such as transportation, urban road layout, electric power and telecommunications, water supply and drainage, landscaping, and public facilities provision, as well as short-term construction (Sit, 1996).

Although the influence of the USSR was important at this time, even beyond the Sino-Soviet split of 1960, it is incumbent upon us to note that there was also a period of "anti-planning" ushered in by Mao Zedong in 1967, linked to the Cultural Revolution. Mao felt that the Beijing Planning Bureau was becoming too strict and bureaucratic (Gu & Cook, 2011, p. 100; Sit, 1996, p. 96), and was limiting the autonomy of the *danwei* or work-unit. These units were thus given free rein within their own spatial jurisdiction, in an ultra-decentralized fashion that was eventually superseded by the need to return to a less anarchic form of urban development, along more traditional socialist lines.

China today is much changed now from that era, and since the reform initiated in the late 1970s, urban planning is contextualized in a different institutional environment. The influence of the Soviet system has faded while open-door policies and market reforms have exerted profound influences on planning. Concurrent with "the third spring" in the planning history since 1949 (Wu, 2007), the style of planning has expanded from blueprint to strategic statement and policy recommendations. Scholars apply different Western theories including local state corporatism, developmental state, entrepreneurship city, and growth coalition to explain the transition of urban governance and planning in Chinese cities (Oi, 1995; Wu, 2002; Xu & Yeh, 2005; Zhu, 1999). Researchers also debate the influences of the changing local political environments on plan making and more recently, implementation (Meligrana, Ren, Zhang, & Anderson, 2008; Ng & Tang, 2004; Tian & Shen, 2011; Wei, 2012). Neoliberal policies such as strategic planning and intercity collaboration were advocated (Wu, 2010). Notions popular in Western planning, such as polycentric development and smart growth, have also been introduced (Yue, Liu, & Fan, 2010).

However, scholars have found that adopting a purely market-oriented or neoliberal perspective to the transformation of planning in China is problematic (Wu, 2010), for this may overlook "path dependence" and "path shaping" inherited from the socialist system in the planning of Chinese cities (Fang & Zhang, 2003; Ng & Tang, 2004; Wang, Shao, Murie, & Cheng, 2012; Wei & Gu, 2010). As Wei (2005) argued, the transition in China is neither market-centric nor state-centric but can be better understood as a triple process of globalization, marketization, and decentralization. The transitional perspective can be elaborated to better understand changes to planning systems in China during the reform period, which also provides a conceptual framework for our study of Beijing.

While the reform has largely dismantled the socialist command economy, marketization and globalization have increased competition and uncertainty, and, empowered by localization, Chinese local states have become more growth oriented and actively engaged in local development and management. On the one hand, market reform and uncertainty do require planning to reflect the invisible hand of the market and change the nature and content of urban planning. Marketization has also made the orthodox approach of socialist physical planning inefficient (Wu, 2007). On the other hand, government and plans are still viewed as indispensable solutions to economic and societal problems. The Chinese government actively makes and implements economic plans, especially the five-year plans, a hallmark of Soviet-style socialism. The central government regulates local development mainly through the evaluation of local political leaders and even the recentralization of local fiscal revenue (Xu & Yeh, 2009). Meanwhile, fiscal decentralization and economic restructuring have provided more incentives for local governments to develop local economies. Urban planning not only responds to reforms and plans at the national level, but has also become a device of local governments to materialize their own development goals, often heavily influenced by local context, which has been conceptualized as growth machine, development state, entrepreneurial state, or transitional institution (Wei, 2005; Wu, 2002; Zhu, 1999). Urban planning in China therefore serves multiple goals and functions, and remains a key instrument of local governments to facilitate economic growth and efficiency, which is often in contrast to planners' interest in justice, equity, participation, and sustainability (Hamnett & Forbes, 2011).

Globalization and China's integration into a global economy have also changed the context and nature of urban planning in China. Master plans in leading Chinese cities have had to increase their global vision of urban development and emphasize the development of a global city as a command and control center of the global economy. Foreign consulting companies and planners have been involved in the process of planning, which is regarded as a function of branding the city (Kipfer & Keil, 2002). Global elements are now mandatory in most master plans in China with a focus on (1) economic globalization to make cities competitive; (2) large-scale infrastructure and urban development projects; (3) global economic networks; and (4) development of services, especially the so-called headquarter economy. The central areas in Chinese cities are largely being transformed toward central business districts (CBDs). With the global push for innovation and sustainable development, notions of sustainability and creative cities have also been introduced to urban planning in China.

In this paper, we argue that the revision of plans in Beijing must be contextualized in the gradual transition of China from a socialist planned economy to an open market economy with a synthesis of globalization, marketization, and decentralization. The interplay between government, investors, and localities has collectively defined the making of plans and their revisions. We also argue that Chinese planning has been facing uncertainty, and the task of planning future urban development is challenging. China's urban planners face the urgent task of adapting traditional master plans to institutional change and economic restructuring. The traditional socialist urban planning system is being dismantled but a new planning system has yet to be fully developed. Consequently, Chinese cities are undergoing many rounds of revision in their master plans, constantly adjusting plans, and facing problems in implementation (Wei, 2005). This process is influenced by "crossing the river by feeling the stones", which can lead to overcautious, incremental, and piecemeal advances in urban planning practices. Last, the reality of urban development also questions the effectiveness of urban planning and points out the insufficient emphasis placed on the assessment and implementation of plans in China (Meligrana et al., 2008; Tian & Shen, 2011). Today, there is often a "back to the future" element, in which Chinese planners seek to adapt global concepts to Chinese situations; hence hybridicity is now an important feature of contemporary Chinese planning.

3. Beijing under Mao: planning a socialist productive city

Beijing is the capital of China and borders Hebei Province and Tianjin Municipality (Figure 1). As the capital, it is sensitive to global contexts, national politics, and development policies; it is also one of the largest and most internationalized cities in China. Its development and planning experiences reflect reforms and transition taking place in Chinese cities. In 2010, the city was composed of 16 districts and two counties, with a total land area of 16,410 km^2. The city can be divided into three zones including the core city, inner suburban, and outer suburban (Feng, Zhou, & Wu, 2008).

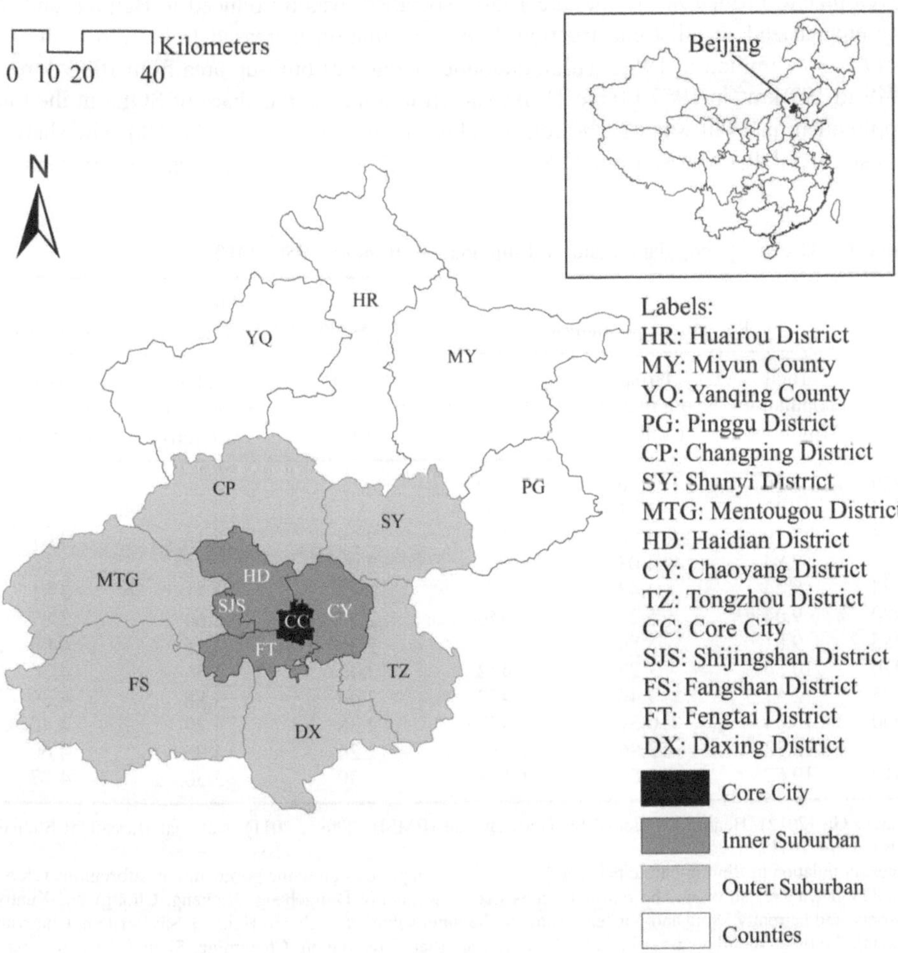

Figure 1. Location and spatial structure of Beijing.

Prior to the modern era, China's cities, including Beijing, were laid out under ancient cosmological principles under the rules of *feng shui* and geomancy. An ancient city with a history of 3,000 years, Beijing ("Northern Capital") became China's capital in the Yuan Dynasty (AD 1271–1368) and has been the capital most of the centuries since, apart from brief periods when Nanjing ("Southern Capital") was the capital. Many historical sites such as the Summer Palace, the Forbidden City, and the nearby Great Wall are key landmarks of the city and China.

After the Opium Wars of the nineteenth century, Beijing was reshaped as a semicolonial city. Western-style buildings, such as embassies, churches, hospitals, banks, and schools were established under what we have called "protomodern" influences, not full-blown modernism because of the feudal remnants of ancient Chinese tradition that meant that nationalist ideas were mitigated against the full adoption of the modernist pantheon of planning concepts from Western countries. These changes nonetheless modified the city image (Hook, 1998).

After the Chinese Communist Party took control of the city in 1949, Beijing underwent radical transformation toward a socialist productive city. Beijing was dominated by state-owned enterprises (SOEs). The city went through three distinctive periods of urban development. From 1949 to the late 1950s, socialism was introduced in Beijing, and the city emphasized socialist construction. Urban population increased from 1.76 million in 1949 to 3.41 million in 1957, with a dramatic increase of built-up area from 108.9 km^2 in 1949 to 192 km^2 in 1957 (Table 1). As shown in Table 2, the share of SOEs in the total employment in 1980 was 82.5%, considerably higher than in 1949 (38.2%). The share of industry in GDP also rose from 32.8% in 1949 to 62.5% in 1980. The spatial expansion of

Table 1. Growth of population and built-up areas in Beijing, 1950–2010.

	Permanent population			Core city	Inner suburban	Outer suburban
Year	Total population (million)	Urban population (million)	Built-up area (km^2)	Total population (million)	Total population (million)	Total population (million)
1949	4.20	1.76	109			
1957	6.33	3.41	192			
1963	7.40	4.56	273	2.28	2.13	3.07
1970	7.84	4.03	–	2.03	2.12	3.56
1975	8.44	4.43	–	2.11	2.31	3.81
1980	9.04	5.21	346	2.34	2.64	3.87
1985	9.81	5.86	–	2.38	3.15	4.05
1990	10.86	7.98	422	2.42	3.56	4.34
1995	12.51	9.46	477	2.44	3.88	4.39
2000	13.64	10.57	488	2.38	4.29	4.40
2005	15.38	12.86	1,182	2.26	4.95	4.60
2010	19.62	16.86	1,186	2.30	5.50	4.77

Source: Gu (2012), Beijing Municipal Statistics Bureau (BMSB, 2009a, 2011), National Bureau of Statistics (NBS, 1994–2011).

Notes: Population in Beijing municipality refers to permanent population while population in subregions refers to population with registration. The core city consisted of originally Dongcheng, Xicheng, Chongwen, Xuanwu districts, and currently Dong and Xicheng districts; the inner suburban included Haidian, Shijingshan, Chaoyang, Fengtai; the outer suburban areas included Daxing, Fangshan, Mentougou, Changping, Shunyi, Miyun, Pinggu, Tongzhou. The built-up area after 2000 includes the built-up area in original rural counties except for Miyun and Yanqing Counties.

Table 2. Socioeconomic indicators of Beijing, 1949–2010.

Year	GDP (billion yuan)	GDP Service sectors (%)	GDP Industrial sectors (%)	Employment State-owned enterprise (%)	Employment Foreign-owned enterprise (%)	Fixed assets investment (billion yuan)	FDI (US$ billion)	Foreign tourists (million)
1949	0.3	40.0	32.8	38.2				
1955	2.4	43.4	31.8	75.0		0.6		
1965	4.0	28.5	52.7	84.6		0.8		
1975	9.1	27.1	61.5	80.7		1.9		
1980	13.9	26.7	62.5	82.5		3.3		
1985	25.7	33.3	50.8	78.5	0.4	9.4		0.9
1990	50.1	38.8	43.8	77.6	1.0	17.9	0.3	1.0
1995	150.8	52.3	35.0	72.7	3.5	84.2	1.4	2.1
2000	316.2	64.8	26.7	58.3	5.2	129.7	2.5	2.8
2005	697.0	69.1	24.8	28.1	7.1	282.7	3.5	3.6
2010	1,411.4	75.1	19.6	20.9	8.3	549.4	6.4	4.9

Source: Beijing Municipal Statistics Bureau (BMSB, 2009a, 2011).

Beijing resulted from the increase of land for the central government and ministries, as well as the postwar recovery and reconstruction (Sit, 1996). The second period from 1958 to 1965 witnessed modest growth in Beijing. The city experienced stagnant economic development, suffering from the "Great Leap Forward" program (1958–1960), natural disasters, and the withdrawal of aid from the USSR. The Great Proletarian Cultural Revolution (1966–1976) and the rural transfer program limited urban growth, during which urban population even dropped from 4.56 million in 1963 to 4.03 million in 1970 and 4.43 million in 1975 (Table 1).

Beijing's master plan followed the Soviet urban planning model and focused on industrial production and building of the socialist capital. The basic concept of planning Beijing was made during 1953 and 1954, with reference to the Moscow Plan of 1935 under Stalin. The socialist style of planning shows the absolute power of the state, and confidence of governments in directing and controlling urban growth. The 1953 master plan projected population growth and built-up area expansion of the city to 5 million and 600 km^2 respectively in 20 years and proposed areas for government, industrial, and educational developments at the center, southeast, and northwest, respectively (Figure 2) (Gu & Cook, 2011). Functional zoning was introduced in which industrial zones were dispersed in suburbs, while new residential areas were located between the old city and the new industrial zones (Sit, 1995). The zones were linked via ring roads and radial subarteries, and separated by greenbelts. Despite the call for the protection of the old city by planners such as Liang Sicheng and Chen Zhanxiang (Dong, 2006), the municipal government preferred to locate the administrative center in the old imperial city, following Russian experiences and Mao's ideas. Since the planning focused heavily on the political function of the city and the socialist ideology, the plan suggested that Beijing should eradicate restrictions of old layouts and patterns.

To make Beijing a socialist productive city, unlike the consumerist emphasis of Western cities, the 1953 Plan privileged industry and production (Table 3). It set up the development goal of Beijing as the political, economic, and cultural center, and the

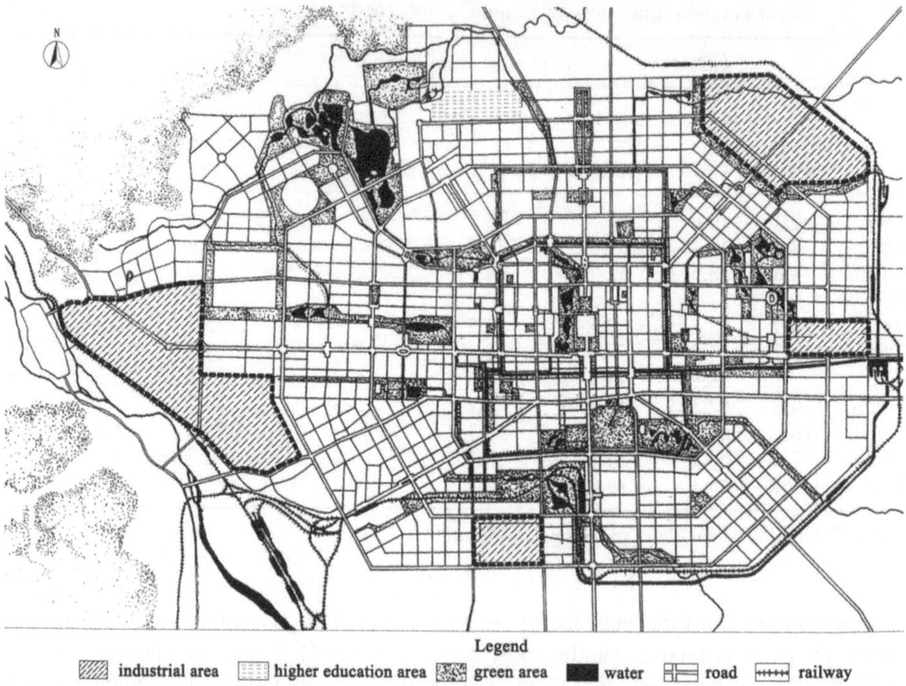

Figure 2. Beijing Master Plan 1954 (last revised version).
Source: Dong (2006, p. 29).

industrial base for the nation. It specified that the capital was for production, for the central government, and for the working people, and that the city should strive to raise the efficiency of labor and production (Gu & Cook, 2011, p. 98). This plan also requested that the existing orthogonal streets be broadened, with the addition of ring roads around the center and radial avenues extending outward, for efficient transportation in all directions. Influenced by the Soviet planning system, urban planning in the 1950s was rooted in the ideology of socialist industrialization and equalitarianism, giving more attention to large-scale industrial projects. Chinese planners also tried to integrate planning approaches such as satellite towns. As in other Chinese cities such as Shanghai, in Beijing a number of satellite towns centered on large-scale SOEs were built up in the suburban areas, where public facilities and residential districts were also constructed. Employees and their families in these SOEs were encouraged to move to these satellite towns. However, the satellite towns were mostly small in size and characterized by the shortage of urban services. People living in these towns still relied heavily on the central city. Consequently, the development of satellite towns in the 1950s had hardly changed the monocentric structure of Beijing.

The 1958 Plan reflects the salient influence of radical changes in political context on the master plan of Beijing (Figure 3). Planning for national political functions and socialist productive cities was central (Table 3). The Soviet concept of a micro-district of 30–60 ha with 10,000–20,000 residents based on limited journeys to work and the formation of working units (*danwei*) was introduced (Gu & Cook, 2011; Lu, 2006). City walls were demolished since they not only restricted the spatial extension of urban development, but also symbolized

Table 3. Beijing Master Plans between 1950 and 2010.

Year of approval or completion	Master plan	Spatial structure	Nature of the city
1953–1954	Beijing Master Plan, 1954	Discrete zones	Center of the whole country, communal living; socialist city; strong industrial, technological, science center
1957–1958	Beijing Master Plan, 1958	Satellite pattern	Political, cultural and education center eliminating the three contradictions; increasingly socialist city serving industrial and agricultural production; rapidly developing into a modern industrial base.
1982	Beijing Master Plan 1982–2000	Dispersed constellation	Renowned, historical, cultural city with hi-tech industry and functioning as the capital of the nation
1993	Beijing Master Plan 1991–2010	Spatial transformation of the east and south; transportation oriented development	Modernized international city of the first rank; hi-tech, tertiary industries
2005	Beijing Master Plan 2004–2020	Two-axis, two-belt, polycentric development	Green city, ecological city, humanistic city, habitable city

Source: Compiled from Sit (1996) and by authors.

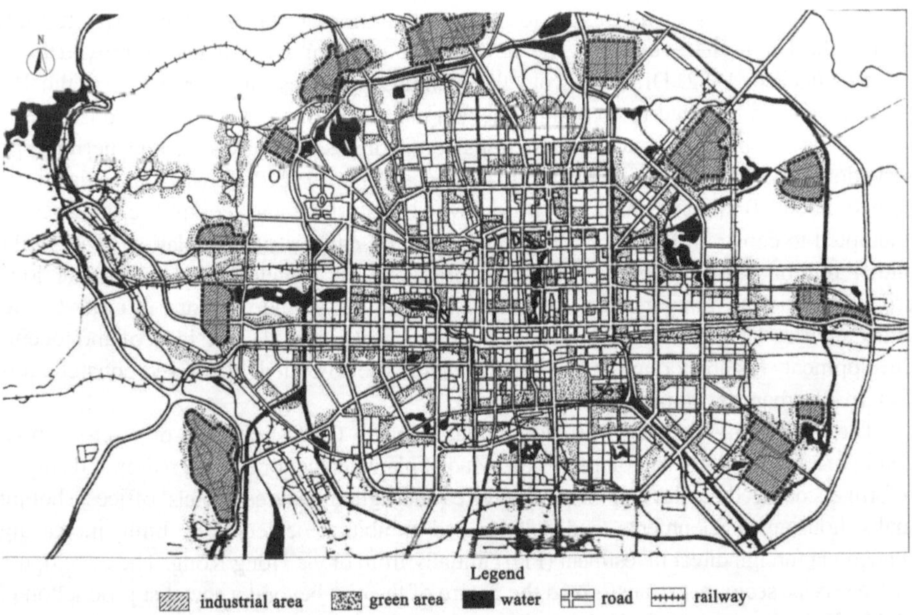

Figure 3. Beijing Master Plan 1958.
Source: Dong (2006, p. 31).

the feudalist, anti-revolutionary order. New construction included the layout of vast boulevards across the city, especially Changan (Heavenly Peace), and the massive expansion of Tiananmen Square to become the focal point of the city and the largest urban square in the world. Ten huge new buildings were built to celebrate the tenth anniversary of the PRC in 1959, including the Great Hall of the People in which the National People's Congress was held. Beijing became a productive city with a focus on heavy industries such as iron and steel and petrochemical production located in inner city areas (Gu & Cook, 2011). However, with respect to citizenship, the city became one of the most austere capital cities in the world, restricting tourists and closing down by 9 pm in the evening. As noted above, the arrival of the Cultural Revolution (1966–1976), meant that in 1967 the city's master plan was suspended and the Beijing Planning Bureau was closed down and not re-established until 1972 (Sit, 1995, p. 205), by which time the worst excesses of the Cultural Revolution were over.

4. Beijing under reform: planning a transitional city

With the death of Mao in 1976, China launched reform in 1978, which triggered economic liberalization and urban development. The urban population in Beijing increased from 4.43 million in 1975 to 5.21 million in 1980 (Table 2). However, the return of urban youth and the rapid population growth also placed a great burden on Beijing, which was suffering from the shortage of urban services and the poor infrastructure that were common to Chinese cities in the early 1980s. In 1980, Beijing had a built-up area of 346 km^2, while the living space per capita was only 4.55 m^2 (China Economy, 2008).

In response to the rising tensions in many largest Chinese cities such as Beijing, the national urban policy in China promoted the development of small- and medium-sized cities and controlled the growth of large cities (Wei, 1994). The central government also directed large cities such as Beijing to initiate urban master plans, which were interrupted during the Cultural Revolution. The new master plan for Beijing was approved by the State Council in 1982. Different from the planning practices during Mao's era, the 1982 Plan recognized the need for economic development and overseas collaborations. Western ideas of urban planning such as functional divisions, satellite towns, and heritage preservation were introduced in planning practices. The 1982 Plan was particularly concerned about Beijing's image as a renowned historical and cultural city. The plan attempted to constrain Beijing's total population and city district population to 10 million and 4 million respectively in 2000, and proposed the development of satellite towns around the city as a major means of population control, similar to plans in other Chinese cities such as Hangzhou (Wei, 2005). As shown in (Figure 4), the idea of monocentric development was abandoned in the 1982 Plan and the so-called "dispersed constellations" pattern of urban development was proposed.

The 1982 Plan, however, was soon outdated due to the fast changes of macroeconomic conditions in the 1980s. In contrast to the Maoist planned economy centered on industry, the reform encouraged the development of services and light industries. Hotels, offices, shopping malls, housing developments, and cultural and exhibition centers were built, increasingly funded via foreign direct investment (FDI) initially from or via Hong Kong. The development of the service sector also transformed the nature of the city beyond a socialist production city (Table 2). Large residential districts, subway extensions, and the second and third ring roads were completed in the late 1980s. The quantity of construction per year for the city rose from 4.5 million m^2 in the late 1970s to 10 million m^2 in the mid- and late 1980s (Zhu, 2009). In

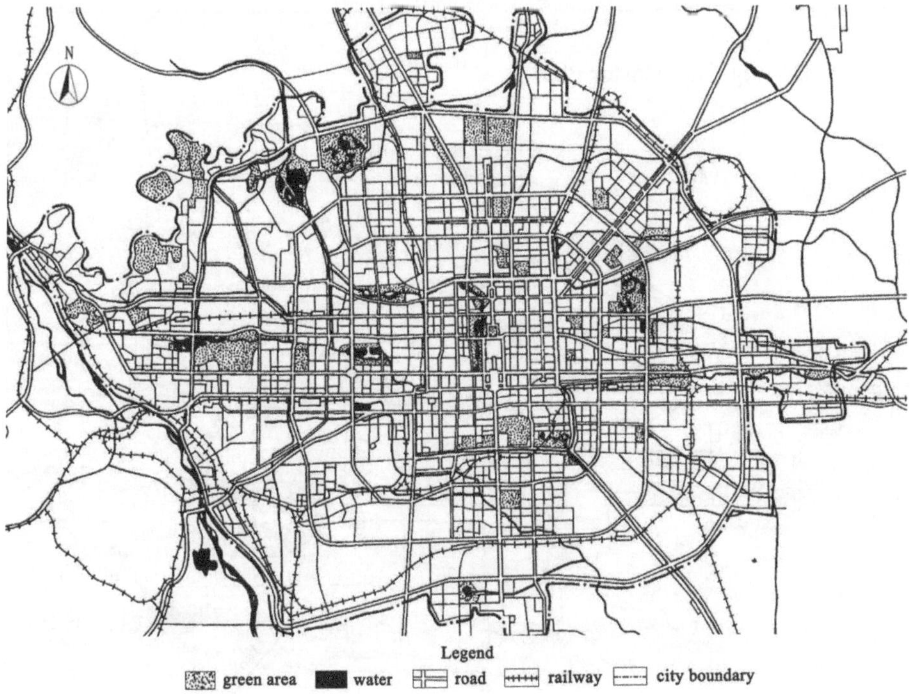

Figure 4. Beijing Master Plan 1982–2000.
Source: Dong (2006, p. 44).

the early 1990s, Deng Xiaoping initiated a new round of economic reforms toward market economies. The market system was formally incorporated into economic development, and many coastal cities were opened up for foreign investment and trade, with the establishment of numerous development zones. Chinese cities experienced rapid economic growth and population increases. These developments in market systems and economy were not anticipated by the 1982 Plan. Also, in the late 1980s, the State Council promulgated the Urban Planning Act (1989) and the Land Management Act (1988). The Urban Planning Act of 1989 was a milestone in the planning history of China, which more clearly defined the responsibility of local planning bureaus for urban development (Gu & Cook, 2011).

5. Making Beijing a global city: recent planning efforts

Beijing City Master Plan for 1991–2010 specified Beijing as an international city, which would be "open in all aspects" and become a "modernized international city of the first rank" in the period from 2010 to 2050 (Gu & Cook, 2011, p. 103). This reaffirmed the need for economic prosperity and the opening up and modernization of China via links with more developed countries. The 1991 Plan sought to increase the GDP of Beijing from 50 billion yuan (US$5.7 billion) in 1991 to 310 billion yuan (US$35 billion) in 2010. The 1991 Plan also promoted the readjustment and distribution of secondary industries, gradually moving them out of the city proper to prevent and control industrial pollution. The plan explicitly emphasized the development of tertiary industries to raise the sector's share in the GDP from 38.8% to 60%. A new CBD was planned and

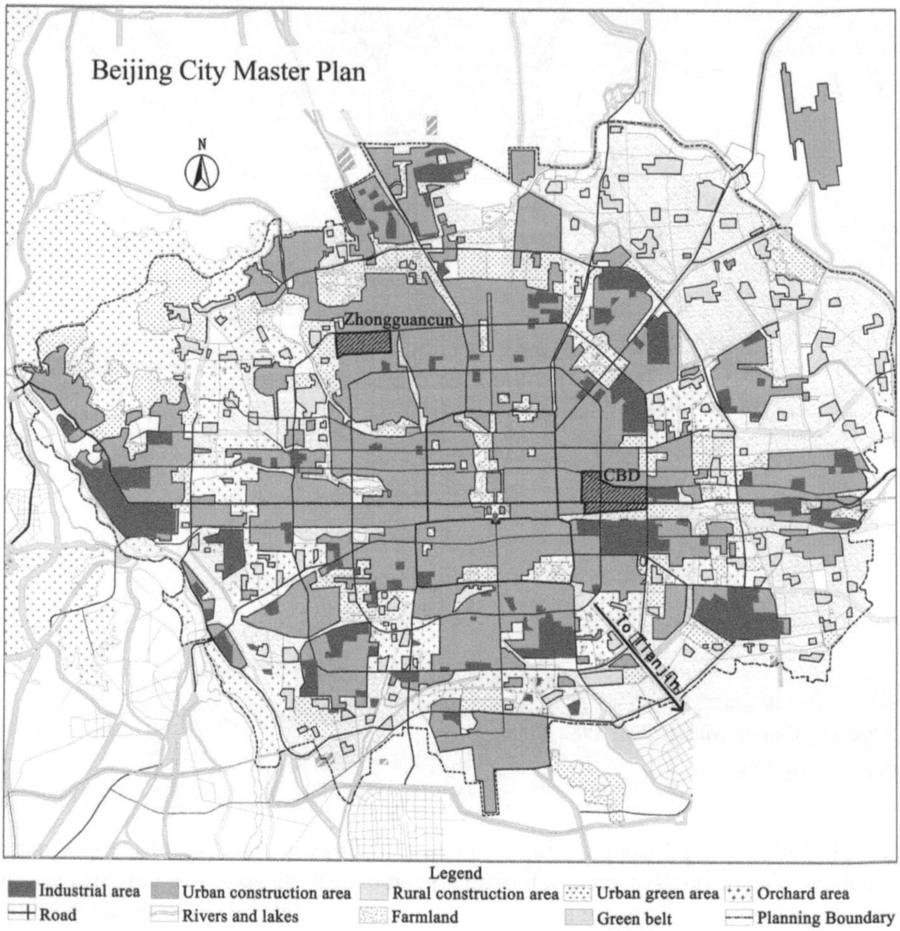

Figure 5. Beijing Master Plan 1991–2010.
Source: People's Government of Beijing Municipality (PGBM, 1991).

constructed in Chaoyangmenwai district (Figure 5). The 1991 Plan also stressed the development of new and high-tech industries, and a number of development zones and high-tech parks were built to support the newly emerging high-tech industries by taking advantage of the richness of human resources in Beijing. Geographical focus was directed to Zhongguanchun, Yizhuan, and Shangdi, with Zhongguancun in particular to be developed as China's Silicon Valley.

With China's accession to the World Trade Organization in 2001, Beijing's economic development has entered a new stage. FDI increased from US$2.5 billion in 2000 to US $6.4 billion in 2010 (Table 2). Beijing has become one of the hotspots of the headquarters and regional centers of global or Fortune 500 enterprises in the world, similar to the concentration of major corporations and financial institutions in the United States (Wheeler, 1986). In 2010, 256 global or Fortune 500 companies had invested in Beijing. The 2008 Olympic Games further enhanced the international reach of Beijing and the number of foreign tourists rose from 2.8 million in 2000 to 4.2 million in 2010. The preparation work for the 2008 Olympic Games also accelerated urban development and investment in infrastructure (Wei & Yu, 2006), with a dramatic increase of fixed asset investment from 129.7 billion yuan in 2000 to 282.7 billion yuan in 2005 and 549.4 billion in 2008 (BMSB 2009b).

The Beijing Action Plan 2002 was made to guide the preparatory work for the Beijing Olympic Games in 2008 (Cook & Miles, 2011). It aimed to push Beijing to "open to the nation and the world in all aspects with international regulations and modern standards". The target was to "raise the level of openness in all aspects of the city of Beijing, and to display to the world a new image of the nation after reform and opening-up" (Gu & Cook, 2011, p. 104). This plan was important as it revealed the building of Beijing as it appeared in 2008. According to the plan, the strategies included the theme of "New Beijing, Great Olympics", the organization principles of "green, science, and technology" and "humanism", and a commitment to showcase images of a "renowned, historical, cultural city", and "a framework of a large, modernized, international metropolis". Infrastructure projects included four subway lines, light rails around the city, a high-speed train to the airport, a second highway to the airport, the fifth and sixth ring roads, and an extension of the Capital Airport. The plan also included the construction of a "social environment" and "modern cultural facilities" such as the Grand National Theatre, China Central Television (CCTV) headquarters, Capital Museum, and the National Museum.

The preparation for the 2008 Olympic Games also triggered planners to formulate a new master plan for 2004–2020 (Figure 6). The 2004 master plan attempts to build Beijing as an ecological, humanistic city, echoing the main themes of Beijing's 2008 Olympic Games. To provide more space for urban development, eight rural counties in the outer suburban area were changed into urban districts in the early 2000s to allow more land for industrial and urban development. The plan proposed to reduce population growth and improve the ecological qualities of the city. The 2004 Plan also emphasized the development of high-tech industries and services, while relocating industry out of the city center and near suburban areas.

The 2004 Plan explicitly divides Beijing into four parts: the core area, development areas, new growth areas, and ecological conservation areas. The ecological conservation areas, including Mentougou, Pinggu, Huairou, Miyun, and Yanqing districts and counties, were set up as ecological belts for Beijing and water resource areas, with a total area of 11,299 km^2, accounting for 68.9% of the total area of the city. The 2004 Plan promoted the concept of polycentric development, with "two axes, two belts and multiple centers (polycentric)" (Gu & Cook, 2011, p. 106). The two axes are running east–west and north–south, crossing at central Beijing. The two belts are the western ecological belt and the eastern development belt. Multiple centers refer to a system of towns and cities, including three major new towns in the suburbs. At Shunyi New Town, a manufacturing base will be developed. A comprehensive service center will be based on Tongzhou New Town; and for Yizhuang, the development objective is to set up a high-tech development center. These new towns are planned to grow into cities with populations between 700,000 and 900,000 by the year 2020.

In 2010, the municipal government of Beijing explicitly advocated the goal of Beijing's development as a full-fledged global city through the development of headquarter economies, technological development, and regional divisions of labor. The cluster development strategy was initiated to foster more extensive global–local production and business linkages, mainly through the development of seven high-end manufacturing and service clusters in the suburban areas (Figure 7). These clusters all focus on the sectors that are regarded as the symbolic-economic elements of a global city, which include an industrial district specializing in green technology in the southwest, three

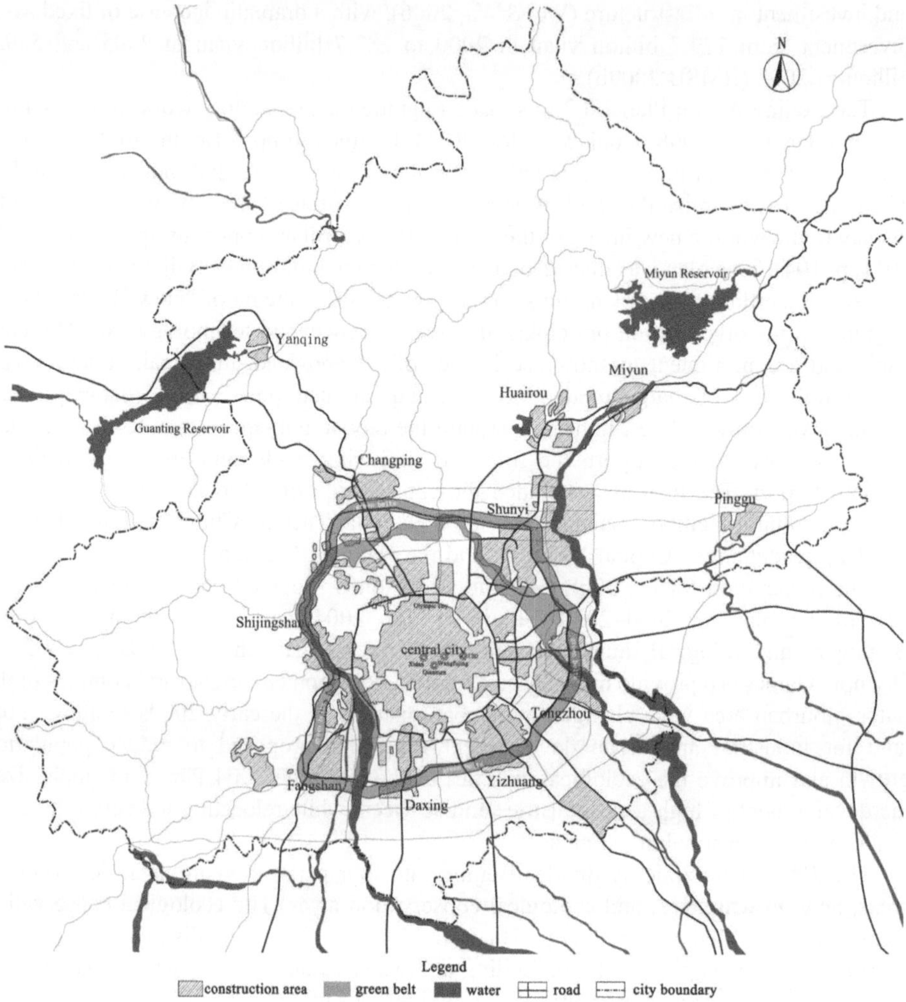

Figure 6. Beijing Master Plan 2004–2020
Source: People's Government of Beijing Municipality (PGBM, 2004).

innovation and advanced manufacturing centers in the west and southeast, a new CBD in the west, and two exhibition and tourism centers in the northeast (Figure 7).

A global city-region approach, which emphasizes the cooperative relationship among different areas and cities in a global city-region, was also employed in the 2010 Beijing Master Plan, which is different from the previous city-system planning approach that focuses on hierarchy and functional divisions. This approach follows Western concepts of urban networking and regional planning that are conceived as wise policy choices for cities to enhance their competitiveness and facilitate regionalization to cope with global economic restructuring (Luo & Shen, 2008). The city-region planning of Beijing also calls for a coordinated development in the greater Beijing area that includes Beijing, part of Hebei Province and Tianjin, another centrally administrated municipality. It also specifies the vision of a global city-region centered on Beijing and emphasizes the importance of

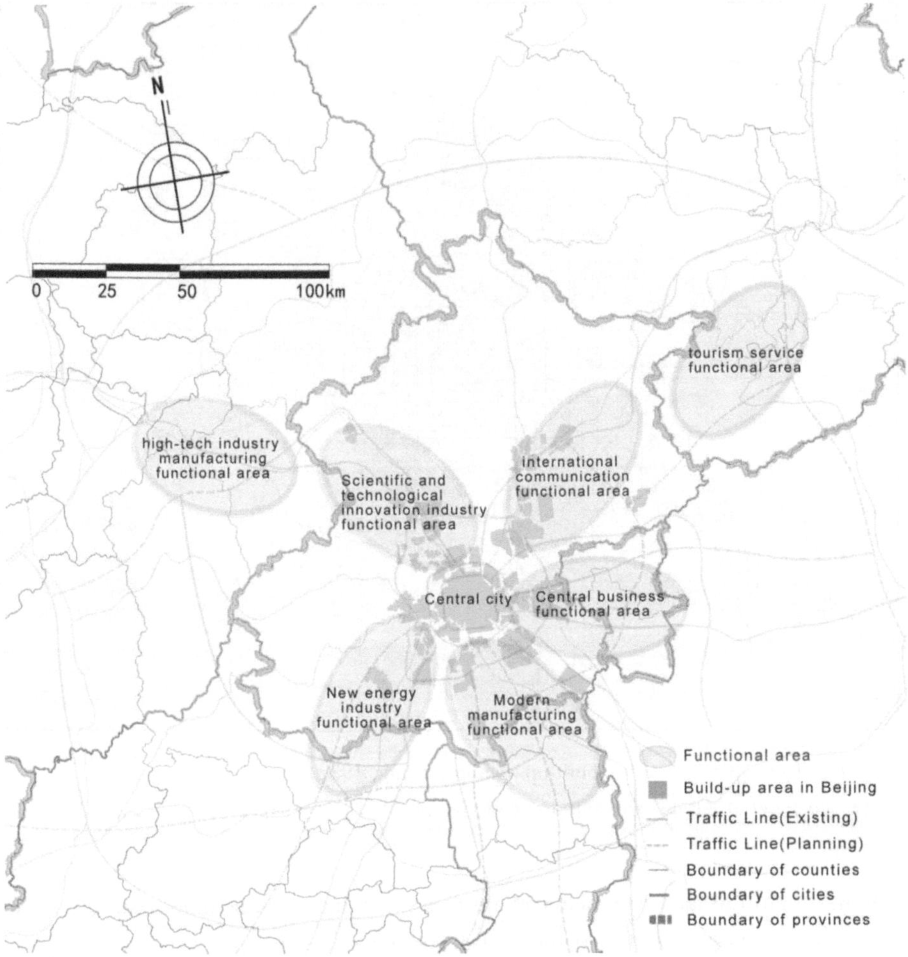

Figure 7. Beijing's vision toward a global city: functional divisions.
Source: Compiled by the authors.

regional cooperation and specialization. As shown in Figure 8, there are three development clusters in this global city-region, including a core subregion centered in Beijing, a coastal development corridor led by Tianjin, and an urban agglomeration in the south of Hebei Province spearheaded by Shijiazhuang, the capital city of Hebei Province.

6. Planning Beijing: vision and reality

Master plans in Beijing have been closely associated with the evolving political-economic context of China's development (Table 3). The visions of master plans were largely transformed from a socialist production city to a global city. Particularly in the reform era, master plans in Beijing attempted to control population growth and increasingly stressed the importance of ecological sustainability, service industries, and livability. However, the process of urban development is a complicated process, and with the

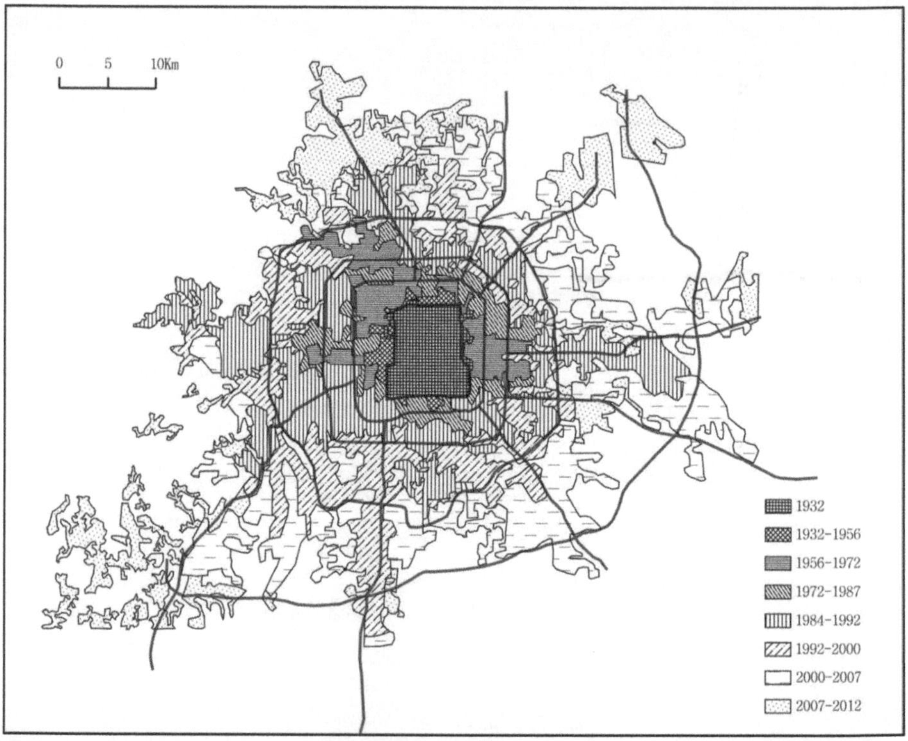

Figure 8. The expansion of the built-up area in Beijing, 1932–2012.
Source: Kuang et al. (2009) and authors' compilation.

declining control of the State, planners are faced with many challenging issues, reflecting mismatches between planning visions and development reality.

First, according to a recent population census in 2010 the population in Beijing reached 19.61 million, which is even larger than the 2004 Plan that sought to limit the population of Beijing in 2020 to 18 million. Polycentric development has not generated job opportunities in satellite towns (Feng et al., 2008), while the spatial mismatch between jobs and housing in the city is another serious problem, eroding livability in the city (Wang, Song, & Xu, 2011). Sprawling development and the concentration of a huge number of youth migrants in the urban fringe brings new challenges for the urban planners in Beijing (Zhao, 2013).

Second, unprecedented urban sprawl is now a serious problem. As shown in Figure 8, most new urban development is concentrated on the urban fringe. The quantity of construction per year increased from 11–12 million m^2 in the 1990s to 20–30 million m^2 after 2000. Urban expansion is accompanied by "zone fever" as Table 4 shows. Even though the city government was cautious in the approval of new development zones (the number of development zones in Beijing dropped from 26 to 19 in 2010), the layout area of these zones actually kept increasing from 131 km^2 in 2002 to 335.74 km^2 in 2010. Most of these development zones were built to attract foreign investment and support industrial development at the local level in suburban areas. Some of these projects compromise long-term environmental sustainability in exchange for short-term economic benefits. Consequently, while the master plans emphasize the conservation of agricultural land, the amount of total

Table 4. Development zones in Beijing, 2002–2010.

Item	2002	2003	2005	2008	2010
Number of development zones	26	27	28	19	19
Layout area of development zones (km^2)	131	166	205	345	336
Accumulated area of land acquisition (km^2)	64	80	109	172	153
Accumulated area of land development (km^2)	58	61	94	133	120
Accumulated area of land development completed (km^2)	55	61	88	122	110
Accumulated number of enterprises	18,921	24,399	30,474	39,031	32,362
Accumulative total investment (billion yuan)	167.5	358.8	536.8	825.5	675.6
Accumulative contracted foreign capital (billion yuan)	127.4	277.7	385.0	643.5	525.9
Foreign funded enterprises (billion yuan)	50.3	85.0	110.0	181.7	162.0
Accumulative contracted foreign capital (US$ billion)	11.6	15.0	11.2	22.8	21.7
Accumulated actual foreign investment (US$ billion)	5.0	10.7	8.1	17.0	15.7

Source: Beijing Municipal Statistics Bureau (BMSB, 2009b, 2011).

arable land in Beijing has experienced a substantial decline from 2,876 km^2 in 2001 to 2,322 km^2 in 2008 (Kuang, Liu, Shao, & Sun, 2009).

Third, unregulated construction in the core area was also criticized. The invasion of massive buildings in the center, breaching of height restrictions, and the destruction of the old courtyard houses (*hutong*) and ancient urban buildings were happening at a greater pace. The Oriental Plaza and the Financial Street, which were planned at 45 m height maximum, reached 68 and 116 m in height, respectively, as the result of aggressive demand for profits by the developers. Although the 2004 Plan aimed to preserve the *hutong* and traditional houses (*siheyuan*), from 2004 to 2008 nearly 162 *hutong* neighborhoods were demolished for new urban developments (China News, 2010). The recent destruction of the former residence of Liang Sicheng, a famous Chinese architect/planner, attracted considerable attention from the public and reminded them of the obstacles that planners in Beijing have to overcome to protect historical sites. The ambiguous development right and insatiable demand for land make planning Chinese cities an even more challenging task.

Fourth, the 2004 master plan also explicitly emphasized the preservation of the ecological area toward an eco-city. Such efforts are important due to the serious challenges for the city in air pollution and natural disasters. Beijing has attempted to reduce resource consumption and to recycle and reuse waste products and to balance social, economic, and natural systems. However, Beijing's sustainability and livability, especially air quality, has deteriorated. This was particularly reflected in a recent survey of local residents on the evaluation of the 2004 master plan. The report shows that local residents are highly dissatisfied with urban planning itself, followed by air quality, the social welfare system, and public services (Table 5).

Specifically, the environment in Beijing suffers from severe natural constraints and lack of water resources. The air quality has become a sensitive topic in Beijing and the central government was reluctant to release data about the density of PM 2.5 pollutants in the air. Many government buildings have their own air filtering systems, while ordinary residents are suffering. In addition, although the 2004 Plan has explicitly addressed the control of cars within 5 million in 2020, the number of cars reached 4.6 million in 2010, which forced the government to randomly select new license numbers to keep the total number under control.

Table 5. Responses of local residents to the 2004 Beijing Master Plan, 2010.

Item	Score	Rank
Social welfare	2.87	10
Community services	2.85	9
Air quality (environment protection)	2.77	8
Housing	2.74	7
Transportation	2.63	6
Municipal-governmental services	2.61	4
Conservation of historic buildings	2.61	4
Government efficiency	2.59	3
Public facilities	2.48	2
City planning	2.36	1
Overall	2.65	

Source: Compiled from Li and Ye (2011).
Notes: A five-point Likert-type scale from 1 (very satisfactory) to 5 (very unsatisfactory) was used to evaluate the implementation of the 2004 master plan. Scores on conservation of historic buildings, city planning and air quality were directly extracted from the questionnaire, and other scores were composite scores based on related questions.

Last, although the aforementioned city-region plan aimed to foster more cooperative relations in the greater Beijing area, the implementation of the plan has been constrained by the complicated administrative framework at the regional level. Politically, the greater Beijing area is composed of one capital city (Beijing), a centrally administered city (Tianjin) and a province (Hebei), as well as a number of prefecture-level and county-level cities or urban districts under their administrations. The implementation of city-region planning has been criticized for overemphasizing the interests of Beijing. For example, many reservoirs in Hebei Province are designated as water sources only for Beijing, even though water shortage is also a key issue in Hebei. In addition, in this relatively affluent region there are 2.7 million people, mainly in Hebei Province, whose incomes are still under the national poverty threshold (Xinhua News, 2006). The development of a competitive global city-region centered on Beijing is therefore undoubtedly overshadowed by the uneven socioeconomic landscape in the region (Yu & Wei, 2008). In these local and regional complications, which impinge upon the "pure" form of urban planning and the global concepts that are being applied, we can see a "hybridicity" similar to that we identified above for the early twentieth century. As indigenous Chinese concepts interweave and interact with global concepts from outside China, then they may become concepts that are no longer purely Chinese, on the one hand, nor purely non-Chinese, on the other hand. This gives rise to hybridicity, which is a fusion of Chinese and global concepts (planning in this case) such that they become transcendent, a synthesis of the Chinese–non-Chinese dialectic. Beijing, in particular, is still a specifically Chinese city, notwithstanding the influence of Western planning concepts, while it also has important general global features.

7. Conclusion

This paper has reviewed the master plans in Beijing and shows that the complexity and rapid change of Chinese cities have necessitated further reform and strengthening of master plan and urban management institutions. The disparity between planning and urban reality is not only attributable to the making of plans and planning systems but also the broad institutional transitions in China. On the one hand, Chinese planners have been slow in adapting to reforms and globalization, and are still widely using traditional, state-centric approaches in urban planning and design. Hence, hybridicity is shown to be at work. Although great improvements have been made since the late 1970s through the increase of personnel, funding, and decision power, current urban planning approaches in China still partially follow the Soviet mode, emphasizing physical planning, project allocation, and urban design, but are weak in social-economic research and market adjustment (Wei, 2005).

As shown in Beijing, the weakness of planning in the prediction of socioeconomic processes is evident. Many types of plans in Chinese cities, such as economic plans, master plans, and land use plans, are operated by different governmental departments and with different time frameworks. Horizontal and vertical fragmentation in management and poor coordination exist among different ministries, departments, bureaus, and their branches within the planning system, which make planning a challenging task (Zhao, 2013). The dynamic contexts of urban planning force revisions between different rounds of planning. The lack of holistic assessment of the previous plans also results in the waste of resource and inefficiency in making plans (Meligrana et al., 2008; Tian & Shen, 2011). Under globalization, Chinese planners do benefit from new planning practices and concepts developed in Western economies (McCann, 2011), although these experiences are selective and confined to China's most globalizing cities. For instance, Western experiences such as public participation and collaborative partnership between the public and private sectors are regarded as unnecessary. Chinese planners tend to follow the ideas of local political elites and seek to "interpret or borrow" Western concepts to decorate their plans (Wu, 2007). Hybridicity is thus a major feature of contemporary urban planning practice in China.

Problems with planning are not only intrinsic in the course of making plans and planning systems, but also the actual implementation. Marketization and decentralization have provided more incentives for local governments to promote local economic development, but also weakened the ability of master plans to coordinate bottom-up urban development. In Beijing, as in other Chinese cities, the control of urban development is still hard to enforce at the local level (Zhao, 2011). The local authorities in urban districts, suburban districts, townships, and villages in the city tend to compromise with investors in exchange for economic growth. The evaluation of the local cadres also depends on the economic indicators such as GDP, foreign investment, and fiscal revenues rather than the preservation of historical sites and agricultural land. Local authorities have few incentives to control urban development and follow master plans (Yang & Wang, 2008). Similarly, as evidenced in Beijing, the recent city-region planning, despite its fancy vision of regional cooperation, has also been facing more challenges in planning implementation given its overemphasizing of the interests of Beijing.

The case of Beijing further implies that while urban planning has had many positive impacts on the growth and management of Chinese cities, considerable efforts are needed to make planning act as an effective guidance for urban development and growth management. More consideration by Chinese planners should be pursued for institutional, social, cultural, and environmental conditions, moving beyond the focus of engineering. More efforts should be made along the lines that improve the transparency and fairness of government policy and encourage people's participation in the planning and decision-making processes, including the participation of international agencies. It would be helpful to have the central government develop better frameworks for planning institutions and planning approaches, with more resources available to build up a more effective Chinese planning structure.

Acknowledgment

We would like to thank comments of the referees, Hongmian Gong, Deborah Martin, and Maureen Clark, and the research assistance of Jing Guo, Xiaohui Yuan, and Haifeng Liao.

Disclosure statement

No potential conflict of interest was reported by the authors.

Funding

We would like to acknowledge the funding of Beijing Philosophy and Social Science Planning Project [grant number 11CSA003] and Chinese Ministry of Education [grant number 11JJDZH005].

References

Abramson, D. B. (2007). The dialectics of urban planning in China. In F. Wu (Ed.), *China's emerging cities: The making of new urbanism* (Chapter 4). London: Routledge.

Beijing Municipal Statistics Bureau. (2009a). *Beijing Liushi Nian* [Beijing 60 years]. Beijing: China Statistics Press.

Beijing Municipal Statistics Bureau. (2009b). *Beijing Tongji Nianjian* [Beijing statistical yearbook]. Beijing: China Statistics Press.

Beijing Municipal Statistics Bureau. (2011). *Beijing Tongji Nianjian* [Beijing statistical yearbook]. Beijing: China Statistics Press.

China Economy. (2008, January 12). *Gaige Kaifang 30nian: Jianzheng Beijing Zhufang 30nian de Bianqian* [30-year in reform: The evaluation of housing in Beijing in the past 30 years]. Retrieved June 9, 2012, from http://bn.ce.cn/djzt/zjsd/200810/31/t20081031_17242975.shtml.

China News. (2010, July 20). Beijing Jiucheng Chaichu Heshi Zhi? Zhengui Wenwu Tieshui xia Xiaoshi [The disappearing historical buildings in Beijing]. Retrieved June 7, 2012, from http://www.chinanews.com/cul/2011/07-20/3197274.shtml.

Cook, I. G. (2006). Beijing as an "internationalized metropolis". In F. Wu (Ed.), *Globalization and China's cities* (Chapter 4). London: Routledge.

Cook, I. G., & Miles, S. (2011). Beijing 2008. In J. R. Gold & M. M. Gold (Eds.), *Olympic cities* (Chapter 17). London: Routledge.

Dong, G. (2006). *Historical capital Beijing: Changes and evolution over 50 years*. Nanjing: Dongnan University Press.

Fang, K., & Zhang, Y. (2003). Plan and market mismatch: Urban redevelopment in Beijing during a period of transition. *Asia Pacific Viewpoint*, *44*, 149–162.

Feng, J., Zhou, Y., & Wu, F. (2008). New trends of suburbanization in Beijing since 1990: From government-led to market-oriented. *Regional Studies*, *42*, 83–99.

Gu, C. L. (Ed.). (2012). *Development plan for Beijing Capital Circle: New approaches on constructing a world city*. Beijing: Science Press.

Gu, C. L., & Cook, I. G. (2011). Beijing: Socialist Chinese capital and new world city. In S. Hamnett & D. Forbes (Eds.), *Planning Asian cities* (pp. 90–130). London: Routledge.

Hall, T., & Hubbard, P. (1998). *The entrepreneurial city*. New York, NY: John Wiley & Sons.

Hamnett, S., & Forbes, D. (Eds.). (2011). *Planning Asian cities*. London: Routledge.

Hook, B. (1998). *Beijing and Tianjin*. New York, NY: Oxford University Press.

Kasarda, J. D., & Parnell, A. M. (1993). *Third world cities*. Newbury Park, CA: Sage.

Kipfer, S., & Keil, R. (2002). Toronto Inc? *Antipode*, *34*, 227–264.

Kuang, W., Liu, J., Shao, Q., & Sun, C. (2009). Spatio-temporal patterns and driving forces of urban expansion in Beijing central city since 1932. *Dili Xinxi Kexue* [*Journal of Geo-Information Science*], *11*(4), 428–435.

Li, D. Q., & Ye, Y. M. (2011). Evaluating Beijing master plan. *Beijing Guihua Jianshe* [Beijing urban planning and construction], *6*, 81–83.

Lu, D. (2006). *Remaking Chinese urban form*. London: Routledge.

Luo, X., & Shen, J. (2008). Why city-region planning does not work well in China: The case of Suzhou–Wuxi–Changzhou. *Cities*, *25*, 207–217.

Ma, L. J. C. (2002). Urban transformation in China, 1949–2000: A review and research agenda. *Environment and Planning A*, *34*(9), 1545–1569.

McCann, E. J. (2011). Urban policy mobilities and global circuits of knowledge: Toward a researchagenda. *Annals of the Association of American Geographers*, 101, 107–130.

Meligrana, J., Ren, W., Zhang, Z., & Anderson, B. (2008). Planning a mega-city's future: An evaluation of Shanghai's municipal land-use plan. *Town Planning Review*, *79*, 267–293.

National Bureau of Statistics. (1994–2011). *China urban statistics yearbook*. Beijing: China Statistics Press.

Ng, M. K., & Tang, W.-S. (2004). Theorising urban planning in a transitional economy: The case of Shenzhen, People's Republic of China. *The Town Planning Review*, *75*, 173–203.

Oi, J. C. (1995). The role of the local state in China's transitional economy. *China Quarterly*, *144*, 1132–1149.

Park, S. O., & Wheeler, J. O. (1983). Industrial location policies and manufacturing employment change: The case of Korea. *Regional Development Dialogue*, *4*, 45–64.

People's Government of Beijing Municipality. (1991). *Beijing Chengshi Zongti Guihua 1991–2010* [Beijing city master plan 1991–2010]. Beijing: Author.

People's Government of Beijing Municipality. (2004). *Beijing Chengshi Zongti Guihua 2004–2020* [Beijing city master plan 2004–2020]. Beijing: Author.

Sit, V. F. S. (1995). *Beijing: The nature and planning of a Chinese capital city*. New York, NY: John Wiley.

Sit, V. F. S. (1996). Soviet influence on urban planning in Beijing, 1949–1991. *The Town Planning Review*, *67*, 457–484.

Tang, W.-S. (2000). Chinese urban planning at fifty: An assessment of the planning theory literature. *Journal of Planning Literature*, *14*, 347–366.

Tang, W.-S. (2006). Planning Beijing strategically: "One world, one dream". *Town Planning Review*, *77*, 257–282.

Tian, L., & Shen, T. (2011). Evaluation of plan implementation in the transitional China: A case of Guangzhou city master plan. *Cities*, *28*, 11–27.

United Nations. (1996). *An urbanizing world: Global report on human settlements*. Oxford: Oxford University Press.

United Nations. (2010). *The state of Asian cities 2010/2011*. Fukuoka, Japan: United Nations Human Settlements Programme.

Walcott, S., & Pannell, C. (2006). Metropolitan spatial dynamics: Shanghai. *Habitat International*, *30*(2), 199–211.

Wang, E., Song, J., & Xu, T. (2011). From "spatial bond" to "spatial mismatch": An assessment of changing jobs-housing relationship in Beijing. *Habitat International*, *35*, 398–409.

Wang, Y. P., Shao, L., Murie, A., & Cheng, J. (2012). The maturation of the neo-liberal housing market in urban China. *Housing Studies*, *27*, 343–359.

Wei, Y. D. (2005). Planning Chinese cities: The limits of transitional institutions. *Urban Geography*, *26*, 200–221.

Wei, Y. D., & Yu, D. (2006). State policy and the globalization of Beijing: Emerging themes. *Habitat International*, *30*, 377–395.

Wei, Y. H. D. (1994). Urban policy, economic policy, and the growth of large cities in China. *Habitat International*, *18*, 53–65.

Wei, Y. H. D. (2012). Restructuring for growth in urban China: Transitional institutions, urban development, and spatial transformation. *Habitat International*, *36*(3), 396–405.

Wei, Y. H. D., & Gu, C. L. (2010). Industrial development and spatial structure in Changzhou city, China: The restructuring of the Sunan model. *Urban Geography*, *31*(3), 321–347.

Wheeler, J. O. (1986). Corporate spatial links with financial institutions: The role of the metropolitan hierarchy. *Annals of the Association of American Geographers*, *76*(2), 262–274.

Wu, F. (2002). China's changing urban governance in the transition towards a more market-oriented economy. *Urban Studies*, *39*, 1071–1093.

Wu, F. (2007). Re-orientation of the city plan: Strategic planning and design competition in China. *Geoforum*, *38*, 379–392.

Wu, F. (2010). How neoliberal is China's reform? The origins of change during transition. *Eurasian Geography and Economics*, *51*, 619–631.

Xie, Y. C., Fang, C., Lin, G. C. S., Gong, H. M., & Qiao, B. (2007). Tempo-spatial patterns of land use changes and urban development in globalizing China: A study of Beijing. *Sensors*, *7*, 2881–2906.

Xinhua News. (2006, October 24). Qidai Dabeijing Guihua rang Jing Ji Shuang Ying [The expectation of "Greater Beijing Plan": A win-win cooperation between Beijing and Hebei province]. Retrieved June 9, 2012, from http://news.xinhuanet.com/fortune/2006-10/24/content_5242942.htm.

Xu, J., & Yeh, A. G. O. (2005). City repositioning and competitiveness building in regional development: New development strategies in Guangzhou, China. *International Journal of Urban and Regional Research*, *29*, 283–308.

Xu, J., & Yeh, A. G. O. (2009). Decoding urban land governance: State reconstruction in contemporary Chinese cities. *Urban Studies*, *46*, 559–581.

Yang, D. Y.-R., & Wang, H.-K. (2008). Dilemmas of local governance under the development zone fever in China: A case study of the Suzhou region. *Urban Studies*, *45*, 1037–1054.

Yu, D., & Wei, Y D. (2008). Spatial data analysis of regional development in Greater Beijing, China, in a GIS environment. *Papers in Regional Science*, *87*, 97–117.

Yue, W., Liu, Y, & Fan, P. (2010). Polycentric urban development: The case of Hangzhou. *Environment and Planning A*, *42*, 563–577.

Zhao, P. (2011). Managing urban growth in a transforming China: Evidence from Beijing. *Land Use Policy*, *28*, 96–109.

Zhao, P. (2013). Too complex to be managed? New trends in peri-urbanisation and its planning in Beijing. *Cities*, *30*, 68–76.

Zhou, Y., & Ma, L. J. C. (2000). Economic restructuring and suburbanization in China. *Urban Geography*, *21*, 205–236.

Zhu, J. (1999). Local growth coalition: The context and implications of China's gradualist urban land reforms. *International Journal of Urban and Regional Research*, *23*, 534–548.

Zhu, J. (2009). *Architecture of modern China*. London: Routledge.

Global cities, cosmopolitanism, and geographies of tolerance

Barney Warf

Department of Geography, University of Kansas, Lawrence, USA

Scholarly attention devoted to global cities has emphasized their economic and political significance. Yet global cities are also deeply cultural and ideological in nature. Contrary to a long tradition of urban studies that views urban life as inherently alienating, this essay argues that the social and cultural diversity of global cities renders them mechanisms for the generation of tolerance and empathy. It opens with a brief review of global cities. Second, it traces the contours of cosmopolitanism as an ideology that fosters respect for social and cultural differences and a wider community of caring. Third, it reviews the empirical evidence of American global cities to make the case that their diversity leads them to generate a progressive cosmopolitan ethnics. It draws on the literature of social psychology, the work of Richard Florida, and indices of immigration, religious and linguistic diversity, homophobia, and voting patterns to substantiate the claim that cosmopolitan values are most evident and deeply rooted in globalized metropolitan centers.

Large cities, and urbanity in general, have long been recognized as centers of cosmopolitanism, that is, as places in which people have access to multiple lifestyles, confront diversity on a daily basis, and frequently adopt a tolerant attitude that accepts social and ethnic differences with a minimum of rancor. The relations between urbanism and tolerance have a long history going back at least to the nineteenth century, when Vaughan (1843/2010), among others, argued that urban life inculcated residents with intelligence and empathetic morality. Because many of their residents tend to swim in deep oceans of information, work in producer services that require relatively high degrees of education, rub shoulders with immigrants, have access to food, music, and movies from around the world, and are exposed to diverse cultures through museums, schools, ethnic festivals, and tourists, large cities—especially global cities—global cities may be seen as repositories of cosmopolitan consciousness. The construction, negotiation, and politics of the geographies of difference and exclusion have become an important line of thought within the discipline of geography (Sibley, 1995; Valentine, 2008), an issue closely related to the spatiality of empathy and caring (Lawson, 2007; Silk, 1998). However, this literature has rarely attempted explicit ties to urban geography. Of particular interest here is the ways in which tolerance, arguably the core cosmopolitan ethic, is produced disproportionately in global cities.

Although it has a lengthy intellectual history, cosmopolitanism is relatively new to academic discourse and to world politics. There are multiple strains within this intellectual tradition that vary by academic discipline and political ideology. Briefly, cosmopolitanism may be defined as an ethical, moral, and political philosophy that uncouples ethical and moral allegiances from geographic location, arguing that each person is bound up with, and obligated to, humanity as a whole (Archibugi, 2004). Cosmopolitans are moral universalists and insist upon the inherent worthiness and dignity of all individuals irrespective of their place of birth. In this view, there are no legitimate grounds for maintaining that some people—fellow nationals, community members, co-religionists—are more worthy than others, that is, those who live far away, are culturally different, or are not constitutive elements of one's self-defined community. The accident of one's birth place is just that—an accident. As Peter Singer (2004) puts it, "Geographical proximity is not in itself of any moral significance." For Daniel Archibugi (2008, p. 143), "Cosmopolitans may be defined as those who know the world and feel at ease anywhere in it." Cheah (1998, p. 22) argues it is "a universal humanism that transcends regional particularism." Martha Nussbaum (1994, p. 1) calls it "a set of loyalties to humanity as a whole."

The connections between global cities, cosmopolitanism, and tolerance are explored here in three stages. First, the paper briefly summarizes global cities and their role within the international division of labor. The focus in this paper is on US global cities, particularly New York and Los Angeles; because cosmopolitanism takes different forms in different contexts, the experience of other global cities may vary considerably. Second, it traces the contours of cosmopolitanism as an ideology that has flourished in the age of rampant globalization. Third, it points to several indices of diversity and tolerance in American global cities that mark them as uniquely cosmopolitan and thus as engines of tolerance. The conclusion ties these themes together and argues that global cities are more than simply engines that produce wealth, information, and power, but also generators of a progressive moral ethics.

It is worth noting that cosmopolitanism has a long history of being generated and embedded in large, internationally oriented urban centers (Holston & Appadurai, 1999; Nava, 2002). Historically, it flourished in trading cities, often those with ports, in which diverse social groups and varying circuits of cultural capital became entangled and intersected (Driessen, 2005). Its emergence, popularity, and occasional decline are contingent and path-dependent, and in some cases cities that once enjoyed a cosmopolitan status, such as those located on the Silk Road, may lose it altogether (Humphrey & Skvirskaja, 2012).

Global cities in context

Contemporary globalization has witnessed the growth of a series of interlinked global cities at the top of the international urban hierarchy. The notion owes much to the pioneering work of John Friedmann (1986). Such places are tied through vast tentacles of investment, trade, migration, and telecommunications to clients and markets, suppliers and competitors, consumers and producers around the world. Global cities serve as the home to massive complexes of financial firms, producer services, and corporate headquarters, or "command and control" centers in the world system. Typically, London, New York, Paris, and Tokyo are positioned at the top (Sassen, 1991/2001, 2006), with cascading layers demarcated by successively smaller roles in the world economy, including Frankfurt, Toronto, Los Angeles, Chicago, San Francisco, Osaka, Hong Kong, Shanghai, Singapore, and Abu Dhabi (Abu-Lughod, 2000; Beaverstock, Smith, & Taylor, 2000). Others have pointed to cities such as Miami, Madrid, Milan, and Beijing. Through their extensive webs of foreign investment, flows of labor, information, commodities, and capital, and through the decision-making powers of the large transnational firms that frequently locate their headquarters there, global cities "fold" global space and time to their advantage.

The core of such conglomerations allow for dense networks of interaction necessary to the performance of headquarters functions, including monitoring frequent changes in niche product markets; negotiating with labor unions; keeping abreast of new technologies and government regulations; keeping an eye on the competition; staying tuned to a complex financial environment; initiating or resisting leveraged buy-outs and hostile takeovers; seeking new investment opportunities, and so forth. The pioneering work of Wheeler and Mitchelson (1989), Mitchelson and Wheeler (1994), and Wheeler (1990) emphasized the role of global cities as corporate headquarters and as hubs in the vast webs of information that link American metropolitan regions together.

The crux of global cities' role in the contemporary world economy is to serve as arenas of interaction, allowing face-to-face contact, political connections, artistic and cultural activities, and elites to rub shoulders easily and exchange tacit knowledge (Abrahamson, 2004). At their core, global cities allow the generation of specialized expertise upon which so much of the world economy depends (Howells, 1990). However, there are important class dimensions to this process: if professionals and elites rely upon face-to-face contact with one another, the day-to-day conditions of the servile service class are often quite different, and their interethnic interactions may be relatively limited or even occasionally hostile in nature (Castaneda, Broadbent, & Coleman, 2010).

Much of the impetus to agglomerate in global cities, despite their manifestly high rents and labor costs, rests on the advantages procured through frequent face-to-face contact, which remains the most efficient and effective means of obtaining and conveying irregular forms of information, particularly when it is highly sensitive in nature. As a sizable body of literature concerned with New York and London has demonstrated, the elites of global cities rely heavily on interpersonal contacts saturated with trust and reciprocity to "get things done" (Longcore & Rees, 1996; Maskell, 1999; Mollenkopf & Castells, 1991). Many such interactions are international in scope, necessitating an appreciation of foreign cultures and languages, as does managing a multiethnic labor force. Assets such as thick networks of interpersonal ties, with all of the issues of trust and external economies of scale that they embody, are very difficult to reproduce in other contexts, an observation that goes far to explain why the primacy of global cities such as Amsterdam, London, and New York has changed very slowly over the last 500 years. The growth of the global telecommunications network has not, contrary to early popular expectations, undermined the status of global cities, but in fact enhanced their stature, particularly when the information produced and exchanged is sensitive, irregular, proprietary, and unstandardized in nature. The Internet and global fiber optics may have increased rather than decreased the need for face-to-face contact: as oceans of ceaselessly circulating data wash over global cities, the interpretive ability to extract meaning from them becomes ever more pressing. For this reason, most managers spend the bulk of their working time engaged in meetings in which the exchange of "buzz," or tacit knowledge, is critical (Storper & Venables, 2004). Indeed, financial and business services firms not only pay high rents to be near city centers and endure the severe congestion such locations often entail, but spend lavishly to fly their executives around the world to meet with their counterparts in person.

Jones (2002) offers an insightful critique of the global cities thesis, noting that it tends to oversimplify the nature of corporate command and control functions by privileging physical location over networks, that is, by focusing on the concentration of head offices.

Rather, he argues, corporate power is wielded throughout the networks of international firms: "to use physical locations as an epistemological framework for theorizing command and control is to a large extent arbitrary and obfuscates the socially constituted complexity of managerial power within the transnational firm" (p. 343). In this light, *all* cities are global cities in that they are inescapably enveloped in worldwide networks of goods, people, capital, and information. Indeed, many smaller cities and even rural areas are also undergoing profound transformations as a result of the information technology revolution.

Global cities are, however, much more than simply nodes of capital accumulation. They are also important centers of discourse and the production of culture. Because their *raison d'être* cannot be reduced to the "economic," but includes a vast variety of formal and informal cultural and political interactions such as tourism, publishing, the printed and electronic media, film and cinema, education, and fashion, global cities are much more than simply poles for the production of corporate knowledge (Knox, 1995). Precisely because they exert so much influence over their respective national economies, and the global one, global cities are also motors of ideological change, absorbing waves of immigrants, socializing people in complex ways around the crisscrossing currents of cultural difference, and generating political agendas that reverberate to shape the possibilities and potentials of action across the globe. In particular, as will be seen, global cities are drivers of a growing cosmopolitan post-national imagination based on tolerance and empathy.

Cosmopolitanism, globalization, and geographies of empathy and tolerance

As a philosophy that emphasizes tolerance and empathy, cosmopolitanism has a long and rich history that arguably can be tracked to classical Greece, particularly to Diogenes, who, when asked where he was from, famously replied "I am a citizen of the world," thus rejecting his local origins and defying the then-prevalent source of identity construction, the *polis* or city-state, in favor of a much larger geopolitical scale, the *cosmopolis*. Similarly, the Stoics suggested that we think of ourselves as surrounded by a series of concentric circles of compassion in which each individual is located at the center of progressively larger webs of mutual obligation extending from the self and family to community to region to the world as a whole, with declining obligations to those farthest removed from ourselves. Cosmopolitanism advocates extending the innermost circles outward, to encompass ever larger domains of humanity.

Immanuel Kant is generally regarded as the foremost Enlightenment thinker of cosmopolitanism. While much of his geography was filled with embarrassing inaccuracies and racist diatribes (Harvey, 2000, 2009), Kant also articulated a clear and coherent vision for a peaceful and tolerant world order. One of the Enlightenment's earliest liberal theorists of international relations, he predicted that the chronic warfare generated by the system of emerging nation-states would eventually entice the world to enter into a peaceful federation founded on international rights that would render war obsolete (1793/1991, p. 90). Following his famous categorical imperative, Kant's 1795 essay, "Towards Perpetual Peace," one of the most popular of his works, emphasizes that this goal is achievable only by treating all human beings as equals. Thus, he argues (1795/1991, p. 107) that "the peoples of the earth have entered in varying degrees into a universal community, and it is developed to the point where a violation of laws in one part of the world is felt everywhere."

Karl Marx, too, was something of a cosmopolitan, as exemplified from the famous quote in the *Manifesto*: "The bourgeoisie has through its exploitation of the world market given a cosmopolitan character to production and consumption in every country" (1848/1973, p. 71). For Marx, cosmopolitanism was much more than an empty, idealistic promise, but an actually existing condition that reflected the extension of capitalist relations of production on a global scale. In this light, Marx regarded nationalism—the ideology that celebrates the nation-state as given, not made, and affirms a geographical imagination that stops at national borders—as akin to religion, viz, false bourgeois consciousness. Whereas Kant's cosmopolitanism was pre-national, Marx's was decidedly post-national in character.

Cosmopolitanism takes as its explicit ethical point of departure human beings, rather than, say, tribes, nations, or religious communities (Pogge, 1992). The cosmopolitan geographical imagination is hence not limited territorially; adherents of cosmopolitanism take as their "imagined community" not the nation-state, but the world. Albert Einstein, advocate of world government, famously described nationalism in 1929 as "an infantile sickness. It is the measles of the human race" (quoted in Dukas & Hoffman, 1979, p. 38). Cosmopolitanism encourages a reflective distance from one's own culture and cultivates a belief in a common humanity (Waldron, 2000). Werbner (2012, p. 153) notes eloquently that: "Because cosmopolitanism is the product of creativity and communication in the context of diversity, it must ultimately be understood ... not merely as individual, but as collective, relational and thus historically situated." Morally, therefore, ethnic, national, religious, and cultural differences are irrelevant and meaningless, or worse, distractions from the important task of caring for others in light of their shared humanity. Compassion knows no borders, for cosmopolitans recognize humanity wherever it occurs: the value of a human life does not vary across space or by social category. For example, this ethic is at play in humanitarian NGOS such as Doctors Without Borders (and numerous similar groups).

The possibilities for the emergence of contemporary cosmopolitanism must be seen in light of contemporary globalization. Although the argument that the nation-state is disappearing has been often advanced in exaggerated and simplistic terms (e.g., Ohmae, 1990), postmodern globalization has fundamentally undermined the territorial order of distinct, mutually exclusive, sovereign states that has underpinned the international order since the Westphalian Peace of 1648. In this light, the nation-state is being eroded simultaneously "from above," that is, by the growing power of transnational institutions (e.g., the IMF, World Bank, and World Trade Organization) and by global financial markets, as well as those working "from below," that is, regional authorities and local movements seeking to attract foreign capital or participate in global processes by bypassing their respective national states as well as NGOs. Globalization has steadily problematized issues of national origin and citizenship, including the gradual emergence of a worldwide civil society (Norris, 2000). Countless entities such as the World Social Forum, international church networks, professional associations, charities and nonprofits, innumerable nongovernmental organizations, environmental groups, human rights advocates, and cross-border trade union activists have gradually gained clout and respectability as some of globalization's most tenacious opponents (Hawken, 2007; Teivainen, 2002). There is, of course, nothing new about transnational solidarities. The abolitionist movement in the nineteenth century, for example, encompassed activists in Britain, Canada, the United States, and elsewhere. Many social movements today struggle against neoliberalism on several fronts, such as the World Social Forum or People's Global Action, often using the Internet to link diverse groups with widely varying agendas together in a strategy Harvey (1996) labels "militant particularism," in which local solidarities find common ground with one another. While these movements are far from constituting a homogenous whole, central to most of their outlooks are tolerance and respect for difference.

Globalization has forced an increasingly widespread recognition that we are all connected to a variety of places worldwide, even if we have never travelled to them. More people are inclusive in their cuisines, movies, music, novels, and fashions they consume, and are more attentive to global events, than ever before. The technologies and institutions that can generate transnational feelings of belonging increasingly play out on a transnational scale, particularly in the age of digital capitalism. Indeed, 33% of the planet's population now uses the Internet, and although there are enormous social and spatial discrepancies in access to it, more people can learn about the rest of the planet more rapidly than ever before. Giddens (1990, p. 187) notes that contemporary everyday life has acquired a global level of extensibility that is historically unprecedented:

> In conditions of late modernity, we live "in the world" in a different sense from previous eras of history. Everyone still continues to lead a local life, and the constraints of the body ensure that all individuals, at every moment, are contextually situated in time and space. Yet the transformations of place, and the intrusion of distance into local activities, combined with the centrality of mediated experience, radically change what "the world" actually is. This is so both on the level of the "phenomenal world" of the individual and the general universe of social activity within which collective social life is enacted. Although everyone lives a local life, phenomenal worlds for the most part are truly global.

In short, just as industrial capitalism induced a "scalar shift" in social processes and identity formation from the local city-state to the nation-state (Giddens, 1985), contemporary globalization is fostering a similar scalar transformation from the national to the global. Peter Singer (2004) notes that in a globalized world the inhabitants of the planet inescapably share one atmosphere, one economy, and, increasingly, one set of common political and ethical concerns. In such a context, the moral community to which each person owes an obligation is, by definition, worldwide, generating an obligation to "care at a distance," in which the concerns of distant strangers are held to be as important as those of people nearby. This transformation makes simple dichotomies such as "us and them," "near and far," "inside" and "outside," and "domestic and foreign" increasingly problematic.

Of course, one may also argue that in some contexts this same process has exacerbated hostilities against ethnic groups, such as the surge of far-right groups in Europe that blame their countries' ills on immigrants. If the primary emotions associated with conservative nationalism are pride and fear of the other, the primary emotions associated with cosmopolitanism are empathy and respect. Nationalism promotes a sense of community, a geography of belonging, centered on ostensibly (and often mythologized) common origins, values, concerns, and expectations, all of which are held to end at the border. Cosmopolitanism, in contrast, seeks to overcome the dichotomous fixation on "we" and "them," citizens and immigrants, the local and the global. As Martha Nussbaum (1994/1996, p. 14) maintains,

> Why should we think of people from China as our fellows the minute they dwell in a certain place, namely the United States, but not when they dwell in a certain other place, namely China? What is it about the national boundary that magically converts people toward whom our education is both incurious and indifferent into people to whom we have duties of mutual respect?

In this view, there is no such thing as an "illegal immigrant," for no human being can be "illegal." Of course, as Tuan (1996) notes, translating the sentiment and attachment to the local is not easily replicable at broader scales, particularly those beyond the nation-state.

Critics of cosmopolitanism argue that its advocates despise the domain of the local as inherently morally inferior and less interesting than the global. Nationalists often deride cosmopolitans as rootless, unattached and uncommitted to the interests, and unconstrained by the restraints of their fellow citizens (e.g., Kofman, 2005). Obviously all forms of belonging inevitably involve local ties as well as nonlocal ones, and no one, of course, can actually be cosmopolitan in the sense of belonging nowhere or everywhere. However, loyalty to humanity need not deprive us of the capacity to care for those who live nearby. "To express solidarity for distant groups of persons does not mean denying it to those who live in our own neighborhood" (Archibugi, 2008, p. 141). Indeed, in the cosmopolitan geographical imagination the local matters every bit as much as do distant places, for cosmopolitans refuse to attach more significance to some people than others (Warf, 2012). In this reading, cosmopolitanism advocates for a continuum of sociospatial attachments, extending the intimate bonds of the family to humanity as a whole, as the Stoics argued two millennia ago.

Muller (2011) prudently cautions against an abstract, simplistic, and utopian stereotype in which metropolitan areas are viewed as the apex of cosmopolitan culture. Rather, the nature and extent of cosmopolitanism among urban populations must be appreciated in grounded, contextualized, empirical terms sensitive to the views of urbanites themselves. Following this line of thought, the next section examines the evidence of cosmopolitanism in the context of American cities.

Are global cities motors of cosmopolitan tolerance? The American evidence

A long tradition of sociological theory, including Durkheim's mechanical and organic solidarity, and Ferdinand Tönnies *Gemeinschaft* and *Gesellschaft*, has portrayed urban life as inherently alienating. This theme was amplified in the early twentieth century by the pioneering urban sociologists and geographers of the Chicago School: Louis Wirth (1938) famously offered a desolate but compelling view of everyday existence in large cities based on their size, density, and heterogeneity. Metropolitan areas in this light were held to be centers of loneliness, desperation, and anomie, in which interpersonal relations were inherently commodified, utilitarian, predatory, and devoid of trust or emotional depth. Conversely, small towns were heralded as fountains of conviviality. This discourse, which has become widely embedded in popular consciousness, leads to widespread but simplistic stereotypes such as the cold, uncaring New Yorker and the friendly, helpful denizens mythologized on television shows such as *Mayberry RFD*.

A cosmopolitan perspective on cities turns this view on its head: here, large cities are seen as places in which the negotiation of difference is perpetual and inescapable, fostering tolerance and acceptance of difference (Binnie, Holloway, Millington, & Young, 2006). Cities have long been synonymous with being sophisticated and worldly, key characteristics of cosmopolitan culture. But they are also arenas in which social and economic success, including at work and in neighborhoods, often necessitates the cultivation of understanding of, and respect for (or at least tolerance of), others who are different. Confronting difference on a daily basis, it is argued, mitigates against the dehumanizing stereotypes that sustain discourses of racism, xenophobia, and nationalism. Such a view is not meant to obscure the intense poverty and inequality concentrated in most urban areas (particularly American ones), or to oversimplify the very real class and ethnic tensions that are often found in metropolitan areas, which have occasionally been sites of racist violence. There is, of course, no inevitable connection between living in culturally diverse environments and becoming more empathetic, tolerant, or cosmopolitan. As Valentine (2008) notes, daily life in such contexts includes as much social exclusion as inclusion (also see Baubock, 2003; Fortier 2007; Ossewaarde, 2007). Indeed, pockets of racism and xenophobia exist within many global city neighborhoods (typically working class communities under economic duress). Competition over jobs can exacerbate interethnic conflicts, such as between African Americans and Koreans in New York (Kim, 2003) or between African Americans and Mexican-Americans in Los Angeles (Waldinger, 1997). Yet a mounting body of evidence suggests that most people living in large cities, particularly global ones, confront the production of social difference in empathetic, often post-national terms.

The relations between social diversity and tolerance are complex, and there is certainly no inevitable link between them. Putnam's influential book *Bowling Alone* (2000) famously asserted that diversity contributed to a withdrawal from civic life, a breakdown in generalized trust, and heightened rampant American individualism. Although his work received widespread popular acclaim, it has been seriously challenged by sociologists, political scientists, and economists on the grounds that his data are not representative, his indices of trust are flawed, and that his analysis is methodologically confused (Anderson & Paskeviciute, 2006; Bjørnskov, 2008; Etzioni, 2001; Laurence, 2011; Leigh, 2006; Letki, 2008; Portes & Vickstrom, 2011; Stolle, Soroka, & Johnston, 2008; Uslaner, 2011).

Mounting evidence from social psychology indicates that diversity leads to tolerance when it challenges stereotypical expectations across multiple domains of lived experience (Crisp & Turner, 2011; Roccas & Amit, 2011). When difference is repeatedly experienced in personal terms in daily life, it becomes difficult to view those of contrasting backgrounds and lifestyles in simplistic, demonized terms (Boisjoly, Duncan, Kremer, Levy, & Eccles, 2006; Gurin, Nagda, & Lopez, 2004; Stephan & Finlay, 1999; Valentine, 2008). For example, people who know gays personally are much less likely to be homophobic; those who work with Muslims realize that claims that many are terrorists are extremely unrealistic and racist. Global cities excel at bringing diverse groups together, fostering opportunities for interethnic and multicultural understanding. Partly for this reason culturally diverse cities tend to perform better economically than less diverse ones (Alesina & Ferrara, 2005; Niebuhr, 2010). Moreover, living in diverse environments allows individuals to forge multiple identities, including in-group membership defined by some dimensions (e.g., sexuality) but not others (e.g., political allegiance). In sociology, diversity is often studied in the context of social capital, that is, the willingness to engage actively in public activities and forge communal linkages (Fieldhouse & Cutts, 2010). Diversity tends to foster numerous cosmopolitan cognitive skills, that is, to make people reconsider their own opinions, to understand the perspective of others, to look past initial dislikes, to accept dissenting ideas, to see more than one side of an issue, and to reduce the drive to force one's own views on others. As Marcus, Sullivan, Theiss-Morse, and Wood (1995, p. 7) put it, "diversity provides an incentive to lessen complete reliance on established beliefs and predispositions." Mutz (2002) demonstrated that people whose personal networks include those with divergent political opinions are more likely to be politically tolerant (although, conversely, it is possible that tolerant people have more friends with different politics). Intergroup contacts reduce prejudice primarily through a process of identification with out-group members (Brown & Hewstone, 2006; Dovidio, Gaertner, & Kawakami, 2003; Pettigrew, 1998; Tafjel & Turner, 1986). Diverse social and cultural environments offer opportunities for interaction in multiple arenas, to cultivate trust, to define mutually beneficial objectives, and to learn to think of different groups as part of a larger, more comprehensive social project.

Conversely, it is most certainly the case that *not* living in a cosmopolitan environment, such as ethnically and religiously homogeneous small towns and rural areas, often leads to a dehumanization, even if subtle, of the perceived Other, usually meaning ethnic, religious, and sexual minorities. A diverse literature from sociology, political science, psychology, and economics confirms this point. Communities lacking cultural diversity often suffer a wet blanket of conformity that acts as an echo chamber in which alternative views and lifestyles are not imagined, or what psychologists call confirmation bias (Nickerson, 1998), leading to exaggerated views of the normality of one's own worldview and lifestyle. Moore and Ovadia (2006) tested the extent of social heterogeneity using census-tract information about the religious, educational, and racial composition of an area and concluded diverse communities tend to be relatively tolerant ones. Indeed, small towns and rural areas frequently exhibit higher rates of racism, religiosity, and nationalistic fervor (Branton & Jones, 2005; Wimmer, 1997). Conversely, living in cosmopolitan environments affords opportunities for pleasant encounters with difference, such as coworkers from different backgrounds or ethnic foods, music, dance, art, and festivals. These statements are most certainly not to be taken as meaning that all small towns are intolerant of diversity, or that all large cities are havens of empathy, but rather that a persistent, if imperfect, pattern exists in which rising city size is associated with greater interaction with, and often respect for, those who are "different."

One indication that cosmopolitan tolerance is concentrated in large cities is found in Richard Florida's (2004, 2005) celebrated notion of the creative class. In this reading, the skilled professionals so central to cognitive-cultural urban economies place a high value on amenities, recreational opportunities, and local social climates that value diversity, including ethnic differences and tolerance of alternative lifestyles (notably gays). Cities that are tolerant, Florida maintains, are economically healthy; those that are restrictive and xenophobic are unlikely to attract the talent they need to remain competitive. Florida and Gates (2004) are adamant that regions hoping to cultivate high technology sectors in particular must foster political contexts marked by social and cultural tolerance (see also Florida, Mellander, & Stolarick, 2008). To illustrate how tolerance ranges across the urban system, Florida (2004) offers the "Tolerance Index," which melds measures of the gay and lesbian population (unmarried same sex partners according to the census), the percent of population that is foreign born, the "bohemian index" (artistic occupations as percent of total), and a measure of racial integration of neighborhoods. The largest cities of the United States generally fall within the lowest quarter of the tolerance ranking (most tolerant), whereas cities under 250,000 people are concentrated in the highest quarter (least tolerant). Thus, using Florida's tolerance index, Seattle comes in at rank 3, followed by Boston (12), Minneapolis (17), San Francisco (20), Los Angeles (32), New York (39), Miami (40), Washington, DC (45), Atlanta (55), Chicago (81), and Philadelphia (85).

Critics of Florida (e.g., Scott, 2006) argue that the creative class thesis underemphasizes the locational dynamic of urban agglomeration and the intense pressure on firms to maximize profits, indeed that it ignores class and celebrates yuppies. Nonetheless, Florida's argument has won wide acclaim. A mounting body of literature has demonstrated that urban diversity is conducive to economic growth, whether in the case of immigrants (Ottaviano & Peri, 2005, 2006), gays (Noland, 2005), or ethnic groups (Quigley, 1998). In this light, tolerance is not simply a matter of being nice to one another, but vital to local economic success.

To illustrate this theme empirically, several measures of the comparative diversity of global cities are now offered that point to tolerance as both an input and output of their status within the global economy. There is, of course, an enormous literature about each of these topics; rather than reiterate them in depth, the following vignettes are designed to highlight the critical role of global cities in the production of a tolerant cosmopolitan consciousness.

Immigration and ethnic diversity

Global cities have long been magnets for immigrants from all over the world, many of whom are trapped in low-paying, unskilled occupations (Wills et al., 2010). Sassen's (1991/2001) seminal work on global cities emphasized that immigrants form an important source of low-wage labor in services and sweat shops, and that their presence both reflects and contributes to the pronounced income polarization found in American global cities. Indeed, despite their cosmopolitanism, large metropolitan areas are also sites of profound class divisions between professionals and a servile service class disproportionately filled with immigrants and people of color, as well as contradictions such as those revolving around gentrification and the corporate reclamation of city centers, as noted by Smith (1995). Some global cities, such as London and Paris, have witnessed violent riots resulting from dispossessed and frustrated immigrants.

Such events serve to remind us that cosmopolitanism is far from a universal condition within such metropolitan areas, but jostles uneasily with other discourses and frames of reference. From the conceptual standpoint of Deleuzian assemblage theory, which conceives of social orders as complex, heterogeneous networks of people, ideas, and places that wax and wane over time, cosmopolitanism and rival ideologies form discursive topologies that fluctuated over time and space, at times setting off intense periods of conflict and interethnic rivalry and at others leading to sustained periods of intercultural cooperation and empathy. Alsayyad and Roy (2006, p. 1) imaginatively capture this notion in arguing that within the globalized city "geographies are constituted through multiple and competing sovereignties" as multiple forms of citizenship are precipitated territorially.

In the United States, New York and Los Angeles, in particular, have long histories of attracting streams of humanity across the planet, and immigrants contribute heavily to their considerable ethnic diversity. In 2010, the five largest US metro areas housed 38% of the country's foreign-born population (Wilson & Singer, 2011), including both legal and undocumented immigrants. In addition to Puerto Ricans, who are US citizens, New York, for example, has large populations of West Indians, Cubans, Russians, Arabs, Chinese, and countless numbers of smaller communities from around the planet. Between 1970 and 2008, the number of non-native New Yorkers doubled to 3 million while the native-born population declined by more than 1 million. Immigrants accounted for 36.4% of New York's population in 2010, twice the percentage of 1970, and 43% of the city's workforce; in Queens, immigrants make up more than half the workforce and in Brooklyn they make up 48%. Similarly, in Los Angeles, the amount of immigrants entering the metro region averaged 800,000 per year from 2000 to 2004, which dropped to about 500,000 per year from 2005 to 2008. Los Angeles has entire neighborhoods dominated by Mexicans, other Latin Americans, Vietnamese, Koreans, Armenians, Filipinos, Chinese, Japanese, and others. Percentages of foreign-born residents among large metropolitan areas include Miami (38.8%), San Jose (36.7%), New York (36.4%), Los Angeles (34.3%),

San Diego (23.5%), Washington, DC (21.8%), and Chicago (21.7%) (Wilson and Singer, 2011).

While this diversity is often seen negatively in light of ostensible "Balkanization," Ellis and Wright (1998) warn that such a perspective foregrounds interethnic tensions and underemphasizes the considerable mixing that such diversity affords. In large part due to the immigrant streams that populate their territories, global cities tend to have relatively smaller shares of White residents than do most other parts of the United States. Whereas non-Hispanic Whites comprised 68% of the US population in 2008, they formed only 53% of the residents of New York City, and relatively small shares of Los Angeles (35%), Chicago (56%), and the San Francisco-Oakland metropolitan region (45%). In contrast, American rural areas tend to hold xenophobic attitudes toward immigration (Fennelly & Federico, 2008).

Linguistic diversity

Exposure to multiple languages on a daily basis is another measure of cosmopolitanism. Based on a study of 44 countries, Anderson and Paskeviciute (2006) show that, contrary to views in which cultural heterogeneity is seen to drive internal conflicts, linguistically diverse civil societies tend to be healthier and more tolerant, including a more politically engaged populace, a willingness to speak directly to others outside of one's in-group, a recognition of the heterogeneity among members of out-groups, and greater involvement in municipal affairs.

The linguistic diversity of eight US global cities is illustrated in Table 1, which reveals that while English is the dominant language in some, it is spoken at home by a bare majority in New York and San Francisco, and a distinct minority in Miami (where non-Hispanic Whites account for only 18.5% of the population). In New York City, if one includes endangered languages spoken by only one or a few people, roughly 800 languages are spoken (Roberts, 2010), leading some to call it the "Multilingual Apple." Almost one-half of all New Yorkers speak a language other than English at home. With 1.1 million students, more than 120 languages are spoken in New York City schools, leading it, like other large cities, to emphasize multicultural educational programs and bilingual instruction. New York City now prints its subway fare cards in nine different languages. Garcia and Fishman (2002, p. 4) note that "English has never been, and cannot be considered today, New York's vernacular."

Table 1. Percent of metro region population over age 5 speaking various languages, 2009.

Metro area	English	Spanish	Other Indo-European	Asian	Other
Atlanta	89.4	5.4	3.1	1.4	0.6
Boston	66.4	14.7	11.3	5.8	1.8
Chicago	65.7	23.3	6.2	3.5	1.3
Los Angeles	39.5	43.8	6.7	8.5	1.5
Miami	22.9	69.4	6.8	0.5	0.3
New York	52.5	23.9	13.3	7.9	2.4
San Francisco	55.8	11.9	6.5	25.3	0.5
Seattle	79.2	4.4	4.1	10.1	2.2

Source: http://www.census.gov/compendia/statab/cats/population.html

Other US global cities are similarly diverse. In Los Angeles County, 135 languages are spoken, and in the San Francisco Bay Area region, 112 (US English Foundation, 2011). Roughly 120 languages are spoken in the Los Angeles County school system, and the school systems of other large cities are comparably diverse, including Boston (140), Chicago (137), and Miami (86).

Religious diversity

The United States is arguably the most religiously diverse country in the world (Eck, 2001) and is far more religious compared to other industrialized countries. However, it has witnessed a gradual decline of mainline Protestant denominations and the growth of numerous smaller faiths, and nonbelievers. Across the face of the country, religious diversity is unevenly distributed (Warf & Winsberg, 2008): in contrast to the essentially monotheistic Bible Belt (Brunn, Webster, & Archer, 2011), in which Baptists are hegemonic, global cities such as New York and Los Angeles exhibit numerous different faiths, as well as higher concentrations of atheists and agnostics. Using several indices of diversity, Warf (2006), using data from the Glenmary Research Center, found that larger urban areas, particularly New York and Los Angeles, exhibited significantly higher degrees of religious diversity than did smaller cities and rural areas. Most minor religions in the United States (i.e., Jews, Muslims, Hindus, and Buddhists) tend to be heavily concentrated in large metropolitan regions. Thus, counties with more than 500,000 residents on average have more than 50 different religions (including denominations), whereas those with less than 100,000 people have less than 20.

Sociological debates about religious diversity focus on its relations to religiosity, that is, the degree to which confronting religious differences affects adherents' participation in religious activities and loyalty to religious institutions. Traditional Weberian interpretations hold that exposure to religious pluralism erodes the naturalness of any given faith and decreases religious adherence (Berger, 1967; Bruce, 2002). Olson and Hadaway (1999, p. 491) argue that "religious pluralism lowers religious involvement by reducing the number of close social ties ... that people maintain with others who share a common religious identity." Not surprisingly, church attendance in large cities tends to be much lower than that outside of the metropolis (Chaves & Stephens, 2003). Many large cities are archipelagos of secularism (Wilford, 2010) in which religious authority has lost most of its appeal. Inasmuch as organized religion is frequently associated with repressive intolerance of difference (as manifested in the religious right), it is fair to assert that the relatively diverse and secularized spaces of globalized metropolises exhibit a certain degree of religious cosmopolitanism not found elsewhere.

Homophobia

A growing literature on geographies of sexuality has demonstrated the importance of place in the construction of sexual identity (Binnie & Valentine, 1999). A specifically queer geography has theorized how space and gay life are intertwined (Brown & Knopp, 2003). With regard to the linkages between urbanity and cosmopolitanism, it is no accident that Florida (2004) found that diverse metropolitan areas exhibit much higher "gay index" (percent of people in same-sex relationships) scores. Florida (2004) maintains that a visible gay and lesbian community is the "canary in the coal mine" of urban tolerance: his "gay index" of same-sex couples living together reveals that larger cities are generally the most tolerant in this regard and puts San Francisco in first place, followed by

Los Angeles (4th), Atlanta (7th), Seattle (8th), Washington, DC (11th), Boston (22nd), and Chicago (24th).

Indeed, the gay rights movement originated in New York (Chauncey, 1994), including the famous Stonewall riots of 1969, and gay clubs are far more common in large cities than in small ones. Many gays and lesbians migrate from small towns and rural areas to large cities to take advantage of the tolerance, anonymity, and open-mindedness there (Watson, 1995). Homophobia has its own moral geography (Leap, 2010), with assaults and hate crimes against gays most likely in small towns (Green, Strolovitch, Wong, & Bailey, 2001; Van Dyke, Soule, & Widom, 2001). More subtly, Michael Brown (2000) has examined how the closet is constructed materially and discursively at multiple spatial scales, and how the metaphor conveys a sense of denial, erasure, and concealment that lies at heart of sexual oppression, a prison that relies on gays to conceal their own identity. Many gays report that it is easier to come out of the closet in large cities than in rural areas. It is no accident that the push to legalize gay marriage has enjoyed its greatest advantages in urbanized, relatively liberal states (Chapman, Leib, & Webster, 2007; Rasmussen, 2006; Webster, Chapman, & Leib, 2010).

Voting patterns

A long-standing feature of American politics is that rural areas and small towns tend to vote more conservatively (i.e., Republican) than do urban areas (Gelman, 2009; Brunn, Webster, Morrill, et al., 2011). Wilson (1985, 1986, 1991) found that the size of one's community is associated with more politically tolerant views. The reasons for this pattern are no doubt complex, but have to do in large part, *inter alia*, with educational levels and the quality and variety of political information available to residents; for example, the conservative television channel Fox News plays an important role in shaping political discourse and voting patterns in rural America. The social ecology of everyday life, viz, the extent to which encounters with difference shape political attitudes, is also critical.

In almost all presidential elections, the proportion of conservative votes drops off steadily as city size increases. For example, in the 2008 presidential election, Obama's votes were disproportionately concentrated in urban archipelagos surrounding large metropolitan areas, such as New York, Boston, Chicago, Atlanta, Denver, Los Angeles, San Francisco, and Seattle, while McCain captured an enormous swath of low-density counties in the West, Midwest, and South. Whereas McCain won decisively in cities with populations less than 100,000, the vote tilted to Obama as city size increased (Table 2). In the 2012 presidential election, Obama won every American city with a population greater

Table 2. Percent of votes for Obama and McCain in 2008 based on city population.

Population (000s)	% Obama	% McCain
<100	44.1	55.9
100–500	57.0	43.0
500–1,000	62.0	38.0
>1,000	64.3	35.7
Total	53.7	46.3

Source: Calculated by author using data obtained from uselectionatlas.org (Leip, n.d.).

than 500,000 by wide margins (average 69%), and his largest absolute margins of victory were generated by New York, Los Angeles, and Chicago; conversely, Mitt Romney won decisively in rural areas and towns with fewer than 50,000 people (Connelly & Marsh, 2012). In part this bias in favor of Democratic candidates reflects the larger concentrations of ethnic minorities and higher educational levels to be found in large cities.

Why do people in larger cities tend to be more politically liberal, and, by implication, cosmopolitan? Clearly the local social context in which voters live matters greatly to their political identity and outlook (Dalton, 2002; Deleon & Naff, 2004). People living in urban areas consistently report higher levels of political tolerance than those living in rural areas (Moore & Ovadia, 2006; Nunn, Crockett, & Williams, 1978; Wilson, 1985, 1991). Political psychologists Park and Peterson (2010) argue that rural, conservative voters tend to be "heart" voters, that is, those who vote on the basis of emotional appeal and in largely personal terms, whereas urban voters tend to be "head" voters, who give more consideration to the impersonal implications of political positions. From a cosmopolitan perspective, living in culturally diverse urban environments, in which voters negotiate social differences on a daily basis, leads them to be more tolerant and politically liberal. Certainly the intense nationalism of the far right has alienated many urban voters, who are more likely to view issues in post-national terms (Jacobs, 2006).

Concluding thoughts

Cosmopolitanism celebrates the commonalities that underlie human life, offering an "imagined community" that extends everywhere. Cosmopolitanism invites a reconfiguration of geographical imaginations, of horizons of belonging. Just as nationalism is a spatial ordering that hinges on borders, othering, and difference, so too is cosmopolitanism a geographic discourse that foregrounds human commonalities and worldwide concerns, and foregrounds values such as sympathy, empathy, and tolerance. By refocusing our scalar gaze, as Isaac Asimov (1970, p. 19) once noted, all the world's issues "will become as intimate as to be one's own." Progressive global politics requires a moral vision suited to the age of contemporary globalization: cosmopolitanism, in confronting the geographies of inequality, offers a route around debilitating nationalism, a means to generate a collective sense of responsibility based on relational ontologies of care and responsibility.

As the empirical examples offered here indicate, global cities tend to be more diverse, tolerant, and cosmopolitan than small ones and rural areas across several dimensions. Florida's Tolerance Index is one such measure of these attitudes, although his work has been criticized for its inattention to class (or this context, the creative class's tolerance for the poor and dispossessed). Global cities are home to diverse ethnic populations, a heterogeneity fueled by immigration. They tend to exhibit enormous linguistic variation. Large cities are more religiously diverse, and more secular, than places lower in the urban hierarchy. They are markedly less homophobic. And not surprisingly, large cities, especially global ones, vote much more liberally than do conservative small towns and rural areas. All of these indices speak to the social and cognitive skill sets that value otherness and refuse to demonize it in simplistic ways. In short, the spaces of global cities, which compel people to interact with those of different cultures, ethnicities, religions, values, nationalities, and sexualities on a daily basis, tend to inculcate their residents with cosmopolitan values that emphasize tolerance, empathy, and respect for difference.

The point here is not either to romanticize global cities nor to demonize small towns and rural areas: not every denizen of small, homogenous communities is an intolerant racist, and not all residents of global cities are open-minded and empathetic. Global cities by no means have a monopoly on the production of cosmopolitan values of empathy and tolerance. Yet the vision of urban life famously promoted by Jane Jacobs (1961)—of metropolitan areas as diverse and bustling centers of creative engagement, of streets in which multiple lifestyles are brought into a creative collision with one another—is most explicitly manifested in those cities with a multitude of connections to the rest of the planet, with people from all corners of the globe, with various social groups living cheek by jowl (Glaeser, 2000). Familiarity, in this case, does not breed contempt, and global cities in this light are repositioned as engines for the production of a progressive morality.

Acknowledgment

The author thanks several reviewers for their careful reading of this paper and useful criticisms.

Disclosure statement

No potential conflict of interest was reported by the author.

References

Abrahamson, Mark (2004). *Global cities*. New York: Oxford University Press.
Abu-Lughod, Janet (2000). *New York, Chicago, Los Angeles: America's global cities*. Minneapolis: University of Minnesota Press.
Alesina, Alberto, & Ferrara, Eliana (2005). Ethnic diversity and economic performance. *Journal of Economic Literature*, *43*(3), 762–800.
Alsayyad, Nezar, & Roy, Ananya (2006). Medieval modernity: On citizenship and urbanism in a global era. *Space and Polity*, *10*(1), 1–20.
Anderson, Christopher, & Paskeviciute, Aida (2006). How ethnic and linguistic heterogeneity influence the prospects for civil society: A comparative study of citizenship behavior. *The Journal of Politics*, *68*(4), 783–802.
Archibugi, Daniel (2004). Cosmopolitan democracy and its critics: A review. *European Journal of International Relations*, *10*(3), 437–473.
Archibugi, Daniel (2008). *The global commonwealth of citizens: Toward cosmopolitan democracy*. Princeton: Princeton University Press.
Asimov, Isaac (1970). The fourth revolution. *Saturday Review October*, *24*, 17–20.
Baubock, Rainier (2003). Reinventing urban citizenship. *Citizenship Studies*, *7*(2), 139–161.
Beaverstock, Jon, Smith, Richard, & Taylor, Peter (2000). World-city network: A new metageography? *Annals of the Association of American Geographers*, *90*, 123–134.
Berger, Peter (1967). *The sacred canopy: Elements of a sociological theory of religion*. New York: Doubleday.
Binnie, Jon, Holloway, Julian, Millington, Steve, & Young, Craig (Eds.). (2006). *Cosmopolitan urbanism*. London: Routledge.
Binnie, Jon, & Valentine, Gill (1999). Geographies of sexuality – A review of progress. *Progress in Human Geography*, *23*(2), 175–187.
Bjørnskov, Christian (2008). Social trust and fractionalization: A possible reinterpretation. *European Sociological Review*, *24*, 271–283.
Boisjoly, J., Duncan, G., Kremer, M., Levy, D., & Eccles, J. (2006). Empathy or antipathy? The impact of diversity. *American Economic Review*, *96*(5), 1890–1905.
Branton, Regina, & Jones, Bradford (2005). Reexamining racial attitudes: The conditional relationship between diversity and socioeconomic environment. *American Journal of Political Science*, *49*(2), 359–372.
Brown, Michael (2000). *Closet space: Geographies of metaphor from the body to the globe*. London: Routledge.

Brown, Michael, & Knopp, Larry (2003). Queer cultural geographies – We're here! We're queer! We're over there, too! In K. Anderson, M. Domosh, S. Pile, & N. Thrift (Eds.), *Handbook of cultural geography*. London: Sage.
Brown, Rupert, & Hewstone, Miles (2006). An integrative theory of intergroup contact. *Advances in Experimental Social Psychology, 37*, 255–343.
Bruce, Stan (2002). *God is dead: Secularization in the West*. Oxford: Blackwell.
Brunn, Stan, Webster, Gerald, & Archer, Clark (2011). The bible belt in a changing South: Shrinking, relocating, and multiple buckles. *Southeastern Geographer, 51*(4), 513–549.
Brunn, Stan, Webster, Gerald, Morrill, Richard, Shelley, Fred, Lavin, Stephen, & Archer, John (Eds.). (2011). *Atlas of the 2008 election*. Lanham, MD: Rowman and Littlefield.
Castaneda, Mario E., Broadbent, M. Ardell, & Coleman, BoaKim (2010). A national overview of the status of inter-ethnic relations in the United States. In Farideh Slal and Rumjahn Hoosain (Eds.), *Democracy and multicultural education* (pp. 73–98). Charlotte, NC: Information Age Publishing.
Chapman, Thomas, Leib, Jonathan, & Webster, Gerald (2007). Race, the creative class, and political geographies of same sex marriage in Georgia. *Southeastern Geographer, 47*, 27–54.
Chauncey, George (1994). *Gay New York: Gender, urban culture, and the making of the gay male world, 1890–1940*. New York: Basic Books.
Chaves, M., & Stephens, L. (2003). Church attendance in the United States. In Michelle Dillon (Ed.), *Handbook of the sociology of religion*. New York: Cambridge University Press.
Cheah, Pheng (1998). The cosmopolitical – Today. In Pheng Cheah & Bruce Robbins (Eds.), *Cosmopolitics: Thinking and feeling beyond the nation* (pp. 20–41). Minneapolis: University of Minnesota Press.
Connelly, Marjorie, & Marsh, Bill (2012, November 11). The building blocks of re-election. *New York Times*, p. 7.
Crisp, Richard, & Turner, Rhiannon (2011). Cognitive adaptation to the experience of social and cultural diversity. *Psychological Bulletin, 137*(2), 242–266.
Dalton, Russell (2002). *Citizen politics: Public opinion and political parties in advanced western democracies* (3rd ed.). Chatham, NJ: Chatham House Publishers.
Deleon, Richard, & Naff, Katherine (2004). Identity politics and local political culture: Some comparative results from the social capital benchmark survey. *Urban Affairs Review, 39*(6), 689–719.
Dovidio, John, Gaertner, Samuel, & Kawakami, Kerry (2003). Intergroup contact: The past, present, and the future. *Group Processes & Intergroup Relations, 6*(1), 5–21.
Driessen, Henk (2005). Mediterranean port cities: Cosmopolitanism reconsidered. *History and Anthropology, 16*(1), 129–141.
Dukas, Helen, & Hoffman, Banesh (1979). *Albert Einstein: The human side*. Princeton: Princeton University Press.
Eck, Diana (2001). *A new religious America: The world's most religiously diverse nation*. San Francisco: HarperCollins.
Ellis, Mark, & Wright, Richard (1998). The Balkanization metaphor in the analysis of U.S. immigration. *Annals of the Association of American Geographers, 88*(4), 686–698.
Etzioni, Amitai (2001). Is Bowling together sociologically lite? *Contemporary Sociology, 30*, 223–224.
Fennelly, Katherine, & Federico, Christopher (2008). Rural residence as a determinant of attitudes toward US immigration policy. *International Migration, 46*(1), 151–190.
Fieldhouse, Edward, & Cutts, David (2010). Does diversity damage social capital? A comparative study of neighbourhood diversity and social capital in the US and Britain. *Canadian Journal of Political Science, 43*, 289–318.
Florida, Richard (2004). *The rise of the creative class*. New York, NY: Basic Books.
Florida, Richard (2005). *Cities and the creative class*. New York, NY: Routledge.
Florida, Richard, & Gates, Gary (2004). Technology and tolerance: The importance of diversity to high technology growth. In Terry Clark (Ed.), *The city as an entertainment machine* (pp. 199–219). New York: JAI Press.
Florida, Richard, Mellander, Charlotta, & Stolarick, Kevin (2008). Inside the black box of regional development: Human capital, the creative class and tolerance. *Journal of Economic Geography, 8* (5), 615–649.

Fortier, Anne-Marie (2007). Too close for comfort: Loving thy neighbour and the management of multicultural intimacies. *Environment and Planning D: Society and Space*, *25*, 104–119.

Friedmann, John (1986). The world city hypothesis. *Development and Change*, *17*(1), 69–83.

Garcia, Ofelia, & Fishman, Joshua (Eds.). (2002). *The multilingual apple: Languages in New York City* (2nd ed.). Berlin: Mouton de Gruyter.

Gelman, Andrew (2009). *Red state, blue state, rich state, poor state: Why Americans vote the way they do*. Princeton: Princeton University Press.

Giddens, Anthony (1985). *The nation-state and violence*. Cambridge: Polity Press.

Giddens, Anthony (1990). *The consequences of modernity*. Cambridge: Polity Press.

Glaeser, Edward (2000). Cities and ethics: An essay for Jane Jacobs. *Journal of Urban Affairs*, *22* (4), 473–493.

Green, Donald, Strolovitch, Dara, Wong, Janelle, & Bailey, Robert (2001). Measuring gay populations and antigay hate crime. *Social Science Quarterly*, *82*(2), 281–296.

Gurin, Patricia, Nagda, Berin, & Lopez, Gretchen (2004). The benefits of diversity in education for democratic citizenship. *Journal of Social Issues*, *60*(1), 17–34.

Harvey, David (1996). *Justice, nature, and the geography of difference*. Oxford: Blackwell.

Harvey, David (2000). Cosmopolitanism and the banality of geographical evils. *Public Culture*, *12* (2), 529–564.

Harvey, David (2009). *Cosmopolitanism and the geographies of freedom*. New York: Columbia University Press.

Hawken, Paul (2007). *Blessed unrest: How the largest social movement in history is restoring grace, justice, and beauty to the world*. New York: Penguin.

Holston, James, & Appadurai, Arjun (1999). *Cities and citizenship*. Durham, NC: Duke University Press.

Howells, J. (1990). The internationalisation of R&D and the development of global research networks. *Regional Studies*, *24*, 495–512.

Humphrey, Caroline, & Skvirskaja, Vera (Eds.). 2012 *Post-cosmopolitan cities: Explorations of urban coexistence*. New York: Berghahn Books.

Jacobs, Anton (2006). The new right, fundamentalism, and nationalism in postmodern America: A marriage of heat and passion. *Social Compass*, *53*, 357–366.

Jacobs, Jane (1961). *The death and life of great American cities*. New York: Random House.

Jones, Andrew (2002). The 'global city' misconceived: The myth of 'global management' in transnational service firms. *Geoforum*, *33*, 335–350.

Kant, Immanuel (1793/1991). On the relationship between theory and practice in international right. In H. Reiss & H. Nisbet (Eds.), *Kant's political writings* (pp. 87–92). Cambridge: Cambridge University Press.

Kant, Immanuel (1795/1991). Towards perpetual peace: A philosophical project. In Hans Reiss & H. Nisbet (Eds.), *Kant: Political writings* (pp. 93–130). Cambridge: Cambridge University Press.

Kim, Claire (2003). *Bitter fruit: The politics of Black-Korean conflict in New York City*. New Haven: Yale University Press.

Knox, Paul (1995). World cities and the organization of global space. In Ron Johnston, Peter Taylor, & Michael Watts (Eds.), *Geographies of global change* (pp. 232–247). Oxford: Blackwell.

Kofman, Eleonore (2005). Figures of the cosmopolitan. *Innovation*, *18*(1), 83–97.

Laurence, James (2011). The effect of ethnic diversity and community disadvantage on social cohesion: A multi-level analysis of social capital and interethnic relations in UK communities. *European Sociological Review*, *27*, 70–89.

Lawson, Victoria (2007). Geographies of care and responsibility. *Annals of the Association of American Geographers*, *97*(1), 1–11.

Leap, William (2010). Homophobia as moral geography. *Gender and Language*, *4*(2), 187–220.

Leigh, Andrew (2006). Trust, inequality and ethnic heterogeneity. *Economic Record*, *82*, 268–280.

Leip, David (n.d.). *Dave Leip's Atlas of U.S. presidential elections*. Retrieved from http://uselectionatlas.org.

Letki, Natalia (2008). Does diversity erode social cohesion? Social capital and race in British neighbourhoods. *Political Studies*, *56*, 99–126.

Longcore, Travis, & Rees, Peter (1996). Information technology and downtown restructuring: The case of New York City's financial district. *Urban Geography*, *17*, 354–372.

Marcus, George, Sullivan, John, Theiss-Morse, Elizabeth, & Wood, Sandra (1995). *With malice toward some: How people make civil liberties judgments*. Cambridge: Cambridge University Press.

Marx, Karl (1848/1973). *Manifesto of the communist party*. London: Harmondsworth.

Maskell, Peter (1999). Globalisation and industrial competitiveness: The process and consequences of ubiquitification. In Edward Malecki & Päivi Oinas (Eds.), *Making connections: Technological learning and regional economic change* (pp. 35–59). Aldershot: Ashgate.

Mitchelson, Ronald, & Wheeler, James (1994). The flow of information in a global economy: The role of the American urban system in 1990. *Annals of the Association of American Geographers*, *84*(1), 87–107.

Mollenkopf, John, & Castells, Manuel (1991). *Dual city: Restructuring New York*. New York: Russell Sage Foundation.

Moore, Laura, & Ovadia, Seth (2006). Accounting for spatial variation in tolerance: The effects of education and religion. *Social Forces*, *84*(4), 2205–2222.

Muller, Floris (2011). Urban alchemy: Performing urban cosmopolitanism in London and Amsterdam. *Urban Studies*, *48*(16), 3415–3431.

Mutz, Diana (2002). The consequences of cross-cutting networks for political participation. *American Journal of Political Science*, *46*(4), 838–855.

Nava, Mica (2002). Cosmopolitan modernity: Everyday imaginaries and the register of difference. *Theory, Culture & Society*, *19*(1–2), 81–99.

Nickerson, Raymond (1998). Confirmation bias: A ubiquitous phenomenon in many guises. *Review of General Psychology*, *2*(2), 175–220.

Nieburh, Annekatrin (2010). Migration and innovation: Does cultural diversity matter for regional R&D activity? *Papers in Regional Science*, *89*(3), 563–585.

Noland, Marcus (2005). Popular attitudes, globalization and risk. *International Finance*, *8*(2), 199–229.

Norris, Pippa (2000). Global governance and cosmopolitan citizens. In Joseph Nye & John Donahue (Eds.), *Governance in a globalizing world* (pp. 155–177). Washington, DC: Brookings Institution Press.

Nunn, Clyde, Crockett, Harry, & Williams, Allen (1978). *Tolerance for nonconformity*. San Francisco: Jossey-Bass.

Nussbaum, Martha (1994/1996). Patriotism and cosmopolitanism. *Boston Review*, *19*(5), 3–34. Reprinted in Martha Nussbaum (Ed.), *For love of country?* (pp. 3–21). Boston, MA: Beacon Press.

Ohmae, Kenichi (1990). *The borderless world: Power and strategy in the interlinked economy*. New York: HarperCollins.

Olson, Daniel, & Hadaway, Christopher (1999). Religious pluralism and affiliation among Canadian counties and cities. *Journal for the Scientific Study of Religion*, *38*(4), 490–508.

Ossewaarde, Marinus (2007). Cosmopolitanism and the society of strangers. *Current Sociology*, *55* (3), 367–388.

Ottaviano, Gianmarco, & Peri, Giovanni (2005). Cities and cultures. *Journal of Urban Economics*, *58*, 304–337.

Ottaviano, Gianmarco, & Peri, Giovanni (2006). The economic value of cultural diversity: Evidence from US cities. *Journal of Economic Geography*, *6*, 9–44.

Park, Nansook, & Peterson, Christopher (2010). Does it matter where we live? The urban psychology of character strengths. *American Psychologist*, *65*(6), 535–547.

Pettigrew, Thomas (1998). Intergroup contact theory. *Annual Review of Psychology*, *49*, 65–85.

Pogge, Thomas (1992). Cosmopolitanism and sovereignty. *Ethics*, *103*(1), 48–75.

Portes, Alejandro, & Vickstrom, Erik (2011). Diversity, social capital, and cohesion. *Annual Review of Sociology*, *37*, 461–479.

Putnam, Robert (2000). *Bowling alone: The collapse and revival of American community*. New York: Simon & Schuster.

Quigley, John (1998). Urban diversity and economic growth. *Journal of Economic Perspectives*, *12*, 127–138.

Rasmussen, Claire (2006). We're no metrosexuals: Identity, place and sexuality in the struggle over gay marriage. *Social & Cultural Geography*, *7*, 807–825.

Roberts, Sam (2010, April 28). Listening to (and saving) the world's languages. *New York Times*. Retrieved from http://www.nytimes.com/2010/04/29/nyregion/29lost.html?pagewanted=all

Roccas, Sonia, & Amit, Adi (2011). Group heterogeneity and tolerance: The moderating role of conservation values. *Journal of Experimental Social Psychology, 47*(5), 898–907.
Sassen, Saskia (1991/2001). *The global city: New York, London, Tokyo*. Princeton, NJ: Princeton University Press.
Sassen, Saskia (2006). *Cities in a world economy* (3rd ed.). Thousand Oaks: Pine Forge Press.
Scott, Allen (2006). Creative cities: Conceptual issues and policy questions. *Journal of Urban Affairs, 28*(1), 1–17.
Sibley, David (1995). *Geographies of exclusion: Society and difference in the west*. London: Routledge.
Silk, John (1998). Caring at a distance. *Ethics, Place and Environment, 1*, 167–182.
Singer, Peter (2004). *One world: The ethics of globalization* (2nd ed.). New Haven: Yale University Press.
Smith, Neil (1995). *The new urban frontier: Gentrification and the revanchist city*. London: Routledge.
Stephan, Walter, & Finlay, Kristina (1999). The role of empathy in improving intergroup relations. *Journal of Social Issues, 55*(4), 729–743.
Stolle, Dietland, Soroka, Stuart, & Johnson, Richard (2008). When does diversity erode trust? Neighborhood diversity, interpersonal trust and the mediating effect of social interactions. *Political Studies, 56*, 57–75.
Storper, Michael, & Venables, Anthony (2004). Buzz: Face to face contact and the urban economy. *Journal of Economic Geography, 4*, 351–370.
Tafjel, Henri, & Turner, John (1986). The social identity theory of intergroup behavior. In William Austin & Stephen Worchel (Eds.), *Psychology of intergroup relations*. Chicago: Nelson-Hall.
Teivainen, Teivo (2002). The world social forum and global democratisation: Learning from Porto Alegre. *Third World Quarterly, 23*(4), 621–632.
Tuan, Yi-Fu (1996). *Cosmos and hearth: A cosmopolite's viewpoint*. Minneapolis: University of Minnesota Press.
US English Foundation. (2011). Retrieved from http://www.usefoundation.org/userdata/file/Research/Regions/california.pdf
Uslaner, Erik (2011). Trust, diversity, and segregation in the United States and the United Kingdom. *Comparative Sociology, 10*, 221–247.
Valentine, Gill (2008). Living with difference: Reflections on geographies of encounter. *Progress in Human Geography, 32*(3), 323–337.
Van Dyke, Nella, Soule, Sarah, & Widom, Rebecca (2001). The politics of hate: Explaining variation in the incidence of anti-gay hate crime. In Betty Dobratz, Lisa Waldner & Timothy Buzzell (Eds.), *The politics of social inequality* (pp. 35–58). Amsterdam: Elsevier.
Vaughan, Robert (1843/2010). *The age of great cities*. London: Woburn Press. Reprinted by Cook Press.
Waldinger, Roger (1997). Black/immigrant competition re-assessed: New evidence from Los Angeles. *Sociological Perspectives, 40*(3), 365–386.
Waldron, Jeremy (2000). What is cosmopolitan? *Journal of Political Philosophy, 8*(2), 227–243.
Warf, Barney (2006). Religious diversity across the North American urban system. *Urban Geography, 27*, 549–566.
Warf, Barney (2012). Nationalism, cosmopolitanism, and geographical imaginations. *Geographical Review, 102*(3), 271–292.
Warf, Barney, & Winsberg, Mort (2008). The geography of religious diversity in the United States. *The Professional Geographer, 60*, 413–424.
Watson, Kath (1995). Get thee to a big city: Sexual imaginary and the great gay migration. *GLQ: A Journal of Lesbian and Gay Studies, 2*(3), 253–277.
Webster, Gerald, Chapman, Thomas, & Leib, Jonathan (2010). Sustaining the "societal and scriptural fence": Cultural, social, and political topographies of same-sex marriage in Alabama. *The Professional Geographer, 62*(2), 211–229.
Werbner, Pnina (2012). Anthropology and the new ethical cosmopolitanism. In G. Delanty (Ed.), *Routledge international handbook of cosmopolitan studies* (pp. 153–166). New York: Routledge.
Wheeler, James (1990). Corporate role of New York City in the metropolitan hierarchy. *Geographical Review, 80*(4), 370–381.

Wheeler, James, & Mitcheson, Ronald (1989). Information flows among major metropolitan areas in the United States. *Annals of the Association of American Geographers*, *79*(4), 523–543.
Wilford, Justin (2010). Sacred archipelagos: Geographies of secularization. *Progress in Human Geography*, *34*(3), 328–348.
Wills, Jane, Datta, Kavita, Evans, Yara, Herbert, Joanna, May, Jon, & McIlwaine, Cathy (2010). *Global cities at work: New migrant divisions of labour*. London: Pluto Press.
Wilson, Jill, and Singer, Audrey (2011). *Immigrants in 2010 metropolitan America: A decade of change*. Washington, DC: Brookings Institution. Retrieved from http://www.brookings.edu/~/media/research/files/papers/2011/10/13%20immigration%20wilson%20singer/1013_immigration_wilson_singer.pdf
Wilson, Thomas (1985). Urbanism and tolerance: A test of some hypotheses drawn from Wirth and Stouffer. *American Sociological Review*, *50*(1), 117–123.
Wilson, Thomas (1986). Community population size and social heterogeneity: An empirical test. *American Journal of Sociology*, *91*(5), 1154–1169.
Wilson, Thomas (1991). Urbanism, migration, and tolerance: A reassessment. *American Sociological Review*, *56*, 117–123.
Wimmer, Andreas (1997). Explaining xenophobia and racism: A critical review of current research approaches. *Ethnic and Racial Studies*, *20*(1), 17–41.
Wirth, Louis (1938). Urbanism as a way of life. In A. Reiss (Ed.), *On cities and social life*. Chicago: University of Chicago Press.

A method for delineating a hierarchically networked structure of urban landscape

Yichun Xie[a] and Ting Ma[b]

[a]Department of Geography and Geology, Eastern Michigan University, Ypsilanti, MI, USA;
[b]Institute of Geographic Sciences and Natural Resources Research, Chinese Academy of Sciences, Beijing, China

We developed a new approach combining statistical and graphical methods to build a hierarchically networked structure for understanding spatial characteristics of urban landscapes at multiple scales. Natural breaks optimization algorithm is applied to determine the optimal number of urban land hierarchies and assign discrete patches into ordered sub-groups according to a selected geometric or functional attribute. Patches contained in a sub-group are linked to the patches in the next sub-group according to the spatial relationships between the patch centroids and Voronoi cells. The conceptual foundations and technical details of this approach are elaborated in the case study of building a hierarchically networked structure of urban built-up patches in Beijing. This approach can be applied to quantify landscape patterns of other land uses to facilitate assessments of interconnection between various types of land uses at varied hierarchical levels (spatial scales) and to evaluate ecological service functions of urban built infrastructures.

1. Introduction

The concept of cities as complex networks has developed rapidly in the last decade as new approaches in complexity science (Neal, 2013; Newman, Barabási, & Watts, 2006; Pijanowski et al., 2014; Portugali, 2010; Tao, Corcoran, Mateo-Babiano, & Rohde, 2014). In *The New Science of Cities* (Batty, 2013), cities are interpreted as complex systems that are growing like biological organisms evolving in a Darwinian fitness landscape. Cities are networked spatial configurations over which different sets of actions, flows, interactions, and transactions are taking place. This type of complexity is better interpreted through urban models built on complexity theories and supported with big data that are being increasingly collected and available in both spatial and temporal dimensions (Szell, 2014). The purpose of this article is to develop a new approach to describe urban landscape as a hierarchically networked system by applying GIS and remote sensing techniques. This new approach integrates both statistical and geometric processes to construct a hierarchically networked system for describing the urban landscape from observed data derived from remote sensing. This research aims at developing a systematic method to measure spatial patterns of urban growth graphically and quantitatively across various regions so that a comparative analysis could be conducted to examine dynamic trajectories and spatial characteristics of urban growth at different stages and sizes.

The study of urban landscapes from the perspective of urban spatial structure and evolution can be traced back to the seminal work of M. R. G. Conzen (1960), a German geographer, and Saverio Muratori (1959), an Italian architect. The school of urban morphology focuses on the analyses of a city's evolution from its founding to present transformation as the outcome of most complex human inventions (Moudon, 1994). The urban form is shaped by social and economic forces over time and also reflects the influences of governmental policies and cultural values (Gong & Wheeler, 2002; Wheeler, 1968, 1973; Whitehand & Larkham, 1992). Urban morphological analyses may involve a wide array of disciplines, including architecture, ecology, economics, environmental science, geography, history, planning, and policy. Valuable reviews of the literatures on urban morphology have been offered by Conzen (1978), Conzen and Whitehand (1981), Whitehand (1992), Franck and Schneekloth (1994), and Moudon (1994, 1997).

One of the noticeable emerging characteristics of urban morphology in modern time is urban sprawl from the perspective of regional planning. Urban sprawl is a common phenomenon in North America, South America, and newly fast growing Asian countries, such as China and India. How to quantitatively measure urban sprawl has been a challenging but hot research topic in economic geography, geostatistics, landscape ecology, and regional planning, and there is a growing body of literature. From the perspective of urban studies and regional planning, regional patterns of urban sprawl can be investigated by density gradient transects (Amson, 1972; Batty & Xie, 1994), kernel density surfaces (Sipper, 1997), the geometry of scatter (Sozio et al., 2013), fractal dimensions (Batty & Xie, 1996), and accessibility measurements (Handy & Niemeier, 1997; O'Sullivan, 2000). In geographic information science (GIS), tempo-spatial analysis of urban patterns is a common theme (Xie & Ye, 2007). Many examples are found in studies of data exploration (Anselin, Syabri, & Kho, 2006; Rey & Janikas, 2006), geostatistics (Goovaerts, 1997; Nielsen & Wendroth, 2003), landscape metrics (McGarigal & Marks, 1995; Xie, Yu, Bai, & Xing, 2006), remote sensing image processing (Xie, Batty, & Zhao, 2007; Xie, Sha, & Bai, 2010), and computational geography (Eppstein, 2005).

Another point worth pointing out is that the concept of landscape ecology constitutes the theoretical foundation of the article and also motivates the analytical solution proposed in this article. A regional landscape is generally considered a discontinuous mosaic of ecological units or an assemblage of patches displaying various community structures and components (Forman & Godron, 1981; Gaucherel, Fleury, Auclair, & Dreyfus, 2006). Although there are slight variations about its definition due to specific purposes (Forman & Godron, 1986; Pickett & White, 1985), the patch-based modeling is commonly regarded as an efficient way to investigate landscape dynamics involving the evolutions of various elements, complicated interactions among them, and their environmental influences (Bogaert, Farina, & Ceulemans, 2005).In specific, a landscape patch implies a closed and internally homogeneous segmentation and represents land surface in a discrete spatial pattern (Kotliar and Wines 1990; Pickett & White, 1985). Despite the fact that such idealized independent spatial entities are rarely observed in the real world, it is reasonably easy to survey landscape patterns through patches because they provide a laconic and intuitive description of complex spatial phenomena and can be efficiently analyzed by applying geo-computation techniques.

In addition, the use of land patches in describing urban land use structure aligns well with the newly developed concept of urban ecosystem service. Urban growth through additions of more built-up urban land patches has affected surrounding areas (peri-urban environments), which disconnect or fragment agricultural and natural landscapes (Bagan & Yamagata, 2012). By considering urban areas as part of a broader ecological system, ecological functions of urban land patches and the impacts on other landscapes from urban built infrastructures can be examined and measured quantitatively (Pataki et al., 2011; Perino, Andrews, Kontoleon, & Bateman, 2014). Knowing this may provide clues pertaining to which alternative urban development options would lead to the best overall environmental outcome.

Furthermore, recent studies have suggested that land patches in the real world often display a hierarchical landscape structure owing to the self-organized behaviors of complex systems (Wu & David, 2002). In other words, spatial patterns of land patches of varying sizes can be depicted as a nested landscape structure at different scales. This new notion of landscape composition has been considered as a paradigm shift in ecology (Gómez, White, & Wulder, 2011; Kotliar & Wiens, 1990; Wu, 2007). As a result, simulations of urban growth and ecological system evolution by using hierarchical patch dynamics have been increasingly reported in urban landscape planning and ecological modeling literature (Sozio et al., 2013; Wu & Loucks, 1995; Zhang, Wu, Grimm, McHale, & Buyantuyev, 2013). However, an easy-to-apply approach to generate hierarchical-structured networks from censored or observed data sets is still desirable for studying patchy landscape. The goal of this article is to develop a new approach to examine spatial patterns of urban landscape in a hierarchical structure by applying GIS and remote sensing techniques. GIS provides a unique toolbox that depicts geographic characteristics of urban landscape so as to reveal complex spatial relationships between features in study and also links quantitative attributes with the geographic features so that, for the first time, physical space can be measured and analyzed in relation to the socio-economic forces that shape it (Moudon, 1997; Wheeler, 1997). In addition, remotely sensed images have proven very valuable mapping both historical and contemporary land use and land cover changes (Xie, Yu, Tian, & Xing, 2005; Xie et al., 2010).

2. Method

2.1. The spatially hierarchical structure of patch objects

Both hierarchy theory and empirical studies have commonly suggested that complex systems often generate hierarchical structures, in which the inner structure of a patch at a given scale is similar to the mosaic structure that contains this patch at a coarser scale (Jorgenson, 2000; Kotliar & Wiens, 1990; Lavorel, Gardner, & O'Neill, 1993). This implies that the patches present at coarser scales can be treated as the dominant patches in comparison with those present at finer scales. Hence, the spatial structure of a patchy landscape can be portrayed as a tree, in which patch objects are organized into different groups and levels according to selected geometric or functional attributes. As shown in Figure 1a, the central site that has the most prominent feature is treated as the top level patch over the entire area. Patches in level-1 group are connected to the top-level patch by the links marked as depth-1. Every site located at a higher level k is itself a local core patch for its neighbor patches at next lower level $k + 1$ and linked with these patches through depth-k links. Therefore, from the viewpoint of urban morphology, a hierarchical structure of patch objects is made of grouped patches and links at different hierarchical levels. Adjacent patches at the same hierarchical level are connected into branches or groups. Patches belonging to higher hierarchical-level groups are also connected

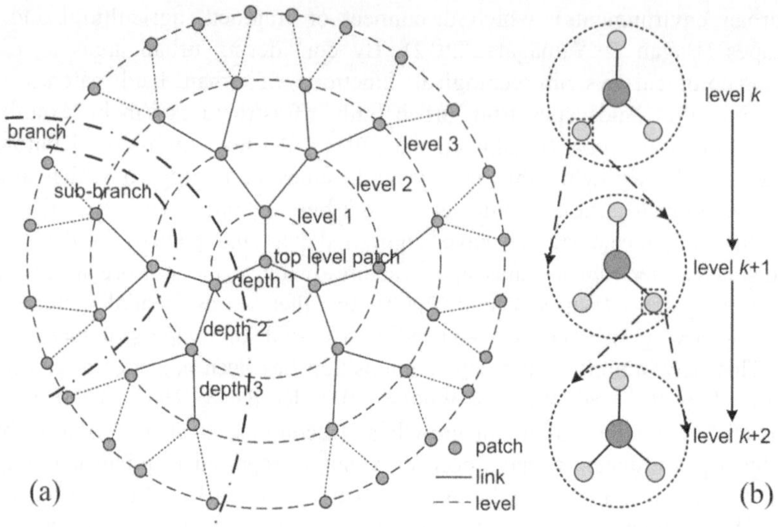

Figure 1. (a) Schematic diagram of the hierarchical structure for patchy landscape and (b) patch-scale changes in the hierarchical structure.

recursively with corresponding lower level groups, in which they act as the core objects of the lower level groups. A local patch cluster dominated by a local core patch at level-k is a branch of another patch cluster at next higher level $k - 1$ (see Figure 1b). Every patch in level-k group becomes a local core patch at next lower level $k + 1$ group.

The aforementioned recursive definition indicates that the hierarchical schema of patch objects is a self-similar structure for scaling measurement. As shown in Figure 1b, darker sites are local core patches, which have variant-level neighbor patches and links in a nested scaling like a network. The core patch could also show distinct behaviors and interactions with other patches located in different level patch clusters. Differing from the strict self-similar system, a hierarchical patch structure of landscape constructed by our approach will show variant branches and sub-branches, because the geometric or functional characteristics and the spatial distribution of patches are rarely uniform. Our hierarchical schema therefore mainly focuses on organizing and clustering patch objects in a series of discrete scales.

2.2. The design of the hierarchical network

It is a challenging task to delineate a hierarchical structure of patchy landscape from a discrete and unordered data set of patch objects, which are obtained through field survey, remotely sensed imagery processing, or other approaches. In general, two procedures are needed in order to construct a hierarchical structure involving patches of varied sizes and with complicated spatial interactions, that is, (1) a clustering method based on a given geometric or functional attribute of patch objects and (2) a spatial partition process. In this study, Jenks natural breaks optimization method is chosen as the statistical method for clustering (grouping) patches, while the spatial partition is performed by the additively weighted Voronoi diagram based on the groups and weights of patch objects. Both Jenks natural breaks method and the Voronoi diagram have been studied extensively in various fields and included in major GIS software packages as standard toolsets. A thorough review of Jenks natural breaks optimization (or classification) method is done by McMaster and McMaster (2002). Comprehensive reviews about Voronoi diagrams can be found in several articles by Srinivasan and Nackman (1987), Gahegan and Lee (2000), and Dong (2008). However, only a couple of applications of Voronoi diagrams in urban landscape studies have been found in the literature (Hall et al. 2001; Tian, Yang, & Xie, 2007). This article is the first effort that integrates Jenks natural breaks classification with Voronoi diagrams to construct a hierarchically networked structure for describing urban morphology.

The general strategy for constructing a hierarchical structure is shown in Figure 2. First, a set of "break values" based on a selected attribute of the patch objects is determined by applying Jenks natural breaks optimization, which maximizes each group's deviation from the means of the other groups. In other words, the patch objects of interest are divided into "a number of groups" by the break values with a descending order of the selected attribute. The patches contained in the top group are considered as the core patches for constructing the first hierarchical level regions according to Christaller's Central Place Theory (Openshaw & Veneris, 2003) The patches included in the second top group are used to construct the next level sub-regions. Sequentially, the study area is spatially subdivided into a series of nested sub-regions, of which the boundaries of local areas are delineated by local core patches using the additively weighted Voronoi diagram. Branches (or local patch clusters) are generated by allocating other patch objects into the corresponding local regions according to spatial relationships. Low-level branches in the

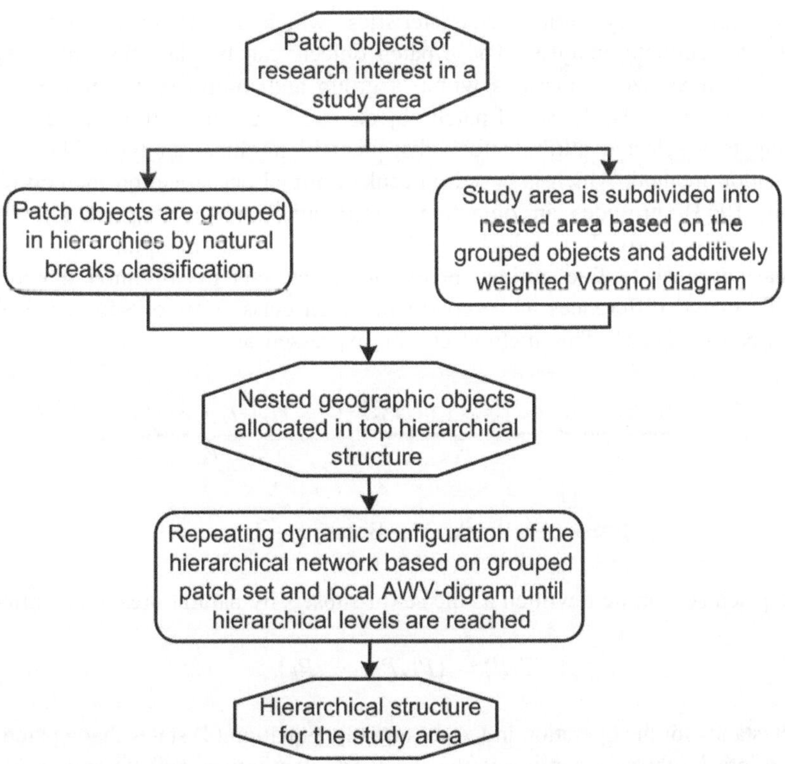

Figure 2. The diagram of the construction of hierarchical structure for patchy landscape.

hierarchical structure are generated by recursively repeating the aforementioned processes in the sub-regions. The details of the procedures will be discussed in the following sections.

2.3. Patch set and natural breaks grouping

A patch object implies a spatially closed and internally connected homogeneous region (Zhang et al., 2013). A patch set is composed of a lot of patch elements with similar land cover properties and variant geometric forms in size and shape or functional features. The patch set represents a kind of land use/cover, which partitions the land surface into a series of clustered but disconnected land pieces. Thus, the patch set P for a given study area can be written as an ordered set of n independent patches according to the specified index in descending order:

$$P = \{p_i \mid i = 1, 2, \cdots, n\} \tag{1}$$

For simplicity, an equivalent radius r (i.e., the radius of a circle that has same area as the corresponding patch does) is adopted to represent the patch. The equivalent circle has the same geographic centroid with its corresponding patch. Thus, P can be expressed as

$$P = \{r_i \mid i = 1, 2, \cdots, n\} \tag{2}$$

Patch objects carry various characteristics, which are described in the form of attributes to facilitate analyses. These patch objects can be classified into a series of groups that maximize similarities within a group and minimize resemblances among groups. In other words, the set of patch objects can be grouped into a series of subsets according to a selected attribute for subsequent hierarchical analysis. Natural breaks classification method, which is based on Jenks optimal classification method (Jenks & Coulson, 1963), provides an objective classification when no prior knowledge is available. This approach is widely applied in choropleth mapping, which offers a statistical approach to find optimal breaks for a data set by minimizing within-class sums of squared differences and maximizing inter-class sums of squared differences (Mennis & Liu, 2005). This method can be expressed as

$$\begin{array}{c} \underbrace{r_1, r_2, \cdots r_g}_{P_1}, \underbrace{r_{g+1}, r_{g+2}, \cdots r_h}_{P_2}, \cdots, \cdots \underbrace{r_{j+1}, r_{j+2}, \cdots r_n}_{P_k} \\ \uparrow \qquad\qquad \uparrow \cdots, \cdots \uparrow \\ \text{Break\_1} \quad \text{Break\_2} \cdots \text{Break\_}k-1 \end{array} \tag{3}$$

The patch set can be rewritten as the set of subsets by natural breaks classification as

$$\Psi(P) = \{P_1, P_2, \cdots, P_k\} \tag{4}$$

where Ψ stands for the operation in Expression (3). Equation (4) states that a patch set P is clustered into k subsets according to the specified quantitative attribute and Jenks natural breaks optimization. The value of k is determined by the researcher through examining the patch set and its characteristics. Because the natural breaks classification runs on an ordered set, the subsets P_i are organized by the descending order.

2.4. *Additively weighted Voronoi diagram partition*

Natural breaks classification provides a statistical method to find out the local dominant patches from an ordered landscape patch set according to the selected attribute. Equation (4) indicates that the patches in the first subset P_1 can be considered as the top dominant patches to generate the first-order spatial partition of the hierarchical structure. A nested structure can be constructed when the process of spatial partitioning is applied repeatedly to the members of next levels of classified patch objects in a top-down order.

Voronoi diagram has been widely employed in spatial decomposition on the basis of a discrete set of points (Edelsbrunner & Seidel, 1986; O'Rourke, 1998), which tend to be applied to explore the sphere of influence of these sites. Ordinary Voronoi diagram (OV diagram) partitions a two-dimensional continuous space into closed sub-regions called Voronoi cells by executing equal-distance segmentation among sites. Let $\|\|$ denote the two-dimensional Euclidean distance of two sites, the boundary of OV cell V formed by sites B_i and B_j can be expressed as

$$V = \{B | \|B - B_i\| = \|B - B_j\|\} \ (i \neq j) \tag{5}$$

Thus, the bisector between two sites constitutes their boundary line. This means that any point located on the boundary should have the same Euclidean or statistical distance to these two sites. AWV diagram in which the distance is multiplied by positive weights is more workable than OV diagram, because it provides a flexible expression for considering the magnitude of sites, which can be computed from any given geometric or functional attributes. Hence, AWV diagram is more suitable than OV diagram when delineating the spatial regions of influence for the patches of various magnitudes. The definition of AWV cell boundary is (Karavelas & Yvinec, 2002)

$$V = \{B | \|B - B_i\| + w_i = \|B - B_j\| + w_j\} \ (i \neq j) \tag{6}$$

where w_i and w_j represent the weights for site B_i and B_j, respectively. In our case study, we only chose the radius of a patch as the weight for the demonstration of building a hierarchical urban structure. The radius is determined by applying the equivalent circle method, which calculates the radius by using the area of the corresponding land patch. This corresponding patch is visualized as a circle with the calculated radius that is placed at the geometric center of the corresponding patch. Practically, the value of weight can be replaced by other quantitative geometric or functional attributes. Equation (6) can be rewritten as

$$V = \{B | \|B - B_i\| + r_i = \|B - B_j\| + r_j\} \ (i \neq j) \tag{7}$$

The geometric shape of AWV cells boundary is hyperbola, which outlines the weighted sphere of patch influences.

If we let Ξ denote AWV operation on a two-dimensional region V, the first-order partition V_1 can be expressed as the function of the first subset P_1 in Equation (4) with the corresponding weight set W as

$$V_1 = \Xi(V, P_1, W) = \{V_i | i = 1, 2, ..., m\} \quad (8)$$

where m is the number of Voronoi cell. The first-order spatial partition of AWV constitutes the top level structure of the hierarchical patches. For any patch r_i, there must exist a Voronoi cell, which contains the centroid of r_i. Thus, every patch can be spatially allocated to its closest cell of AWV diagram according to the spatial relationship between the centroid and the boundary of Voronoi cell. If we use Θ to represent the patch allocation, this process can be expressed as

$$P^1 = \Theta(P_1, V_0) = \left\{P_1^1, P_2^1, \cdots, P_g^1\right\} \quad (9)$$

where the superscript "1" represents the first-order partition. In order to obtain the multilevel hierarchical structure of a patchy landscape, these processes including AWV diagram partition should be executed iteratively as follows:

(1) Allocate the patches in the first-order patch group (class) P_k into the hierarchical level "1" Voronoi cells according to the spatial relationship

$$P^k = \Theta(P_k, V_k) \quad (10)$$

(2) Partition the Voronoi cells of $V_{k=1}$ by applying AWV diagram and the selected weight:

$$V_k = \Xi(V_{k=1}, P^k, W) \quad (11)$$

(3) Allocate the patches in the second group and partition the hierarchical level "2" Voronoi cells by following similar processes as (1) and (2) (Note: the partitions of the hierarchical level 2 Voronoi cells are completely within the partitions of the level 1 Voronoi cells, forming a nested structure).
(4) Connect the first-order patches in the hierarchical level 1 partitions with the corresponding second-order patches contained in the nested hierarchical 2 partitions to construct the hierarchical level 1 network.
(5) Repeat the aforementioned four processes sequentially to build the hierarchically networked urban patch system and terminate the allocating, partitioning, and connecting processes when the preset hierarchical level is reached.

A hierarchical spatial partition and its corresponding networked organization of patchy landscape is obtained through the above algorithm. However, in order to describe or extract quantitative information of the hierarchical structure, we need to design some indices to indicate the urban network characteristics and landscape dynamics (see Section 2.5).

2.5. *Statistical indices for the hierarchical structure*

We introduce three statistical indices—connectivity, accessibility, and uncertainty—to quantify the hierarchical structure of patchy landscape derived from the Graph Theory (Diestel, 2005) and urban network analysis (Batty, 2013). The connectivity C is taken as the number of patches which are found in a Voronoi cell and connected to the local core patch at a higher hierarchical level. The accessibility index A portrays the total effect of all other patches upon the local core patch within a Voronoi cell. It is computed as a function of the patch magnitude a_j and the spatial distance d_j between the local core patch and other patches as

$$A = \sum \frac{a_j}{d_j} \tag{12}$$

The uncertainty index U, derived from Shannon's entropy, is used to measure the diversity of patch cluster at a given hierarchical level within a Voronoi cell. The definition of U is

$$U = -\sum \left(\frac{a_i}{A}\right) \log\left(\frac{a_i}{A}\right) \tag{13}$$

where a_i is the magnitude of a patch and A is the total magnitude of all patches connected to a common local core patch at a given hierarchical level. A reaches its maximum value when all patches have the same magnitude.

3. Case studies

3.1. Study area and data preprocessing

Located in north China, Beijing consists of 16 districts and two counties with an area of nearly 16,000 km^2. As one of rapidly growing cities in China, the landscape of Beijing has witnessed dramatic changes in the recent decades (Ding, 2004; Gaubatz, 1999). Spatially uneven development of human settlements could lead to a hierarchical structured network of urban patches (Antrop, 2000; Fujita, Krugman, & Mori, 1999). In this section, we attempt to apply the aforementioned method to demonstrate how to construct a hierarchically networked structure of urban landscape and to apply the statistical indices discussed in Section 2.5 to examine the characteristics of the hierarchical structure.

In order to extract a hierarchical structure, the remote-sensed image obtained by Landsat Thematic Mapper in 2004 is used to identify the urban and build-up patches by a supervised classification method. Because our method is based on the two-dimensional vector object, urban and build-up patches are converted to the polygon objects as shown in (Figure 3a). A total of 1557 urban and build-up patches are identified, and the corresponding ordered patch sets are generated by the equivalent circle method (as shown in (Figure 3b) for subsequent analyses.

3.2. The hierarchical structure of Beijing City

The areas of the urban and build-up patches are chosen to be the quantitative weights to create a five-level hierarchical structure of the patchy landscape of Beijing. The choice of five hierarchical levels is primarily based on the administrative structure of Beijing: the municipality—central business district, administrative districts, counties, towns, and villages. In addition, the selection of five groups (classes) reflects the size distribution of 1557 urban and build-up patches in terms of the numbers of patches falling within each hierarchical level. Hence, the original ordered patch sets are clustered into five sub-groups (representing five hierarchical levels) by using the natural breaks classification method. Because the urban region of Beijing contains a prominent core patch that is over 50% of the total urbanized land, there is only one patch object in the first-order sub-group. The second sub-group is composed of six

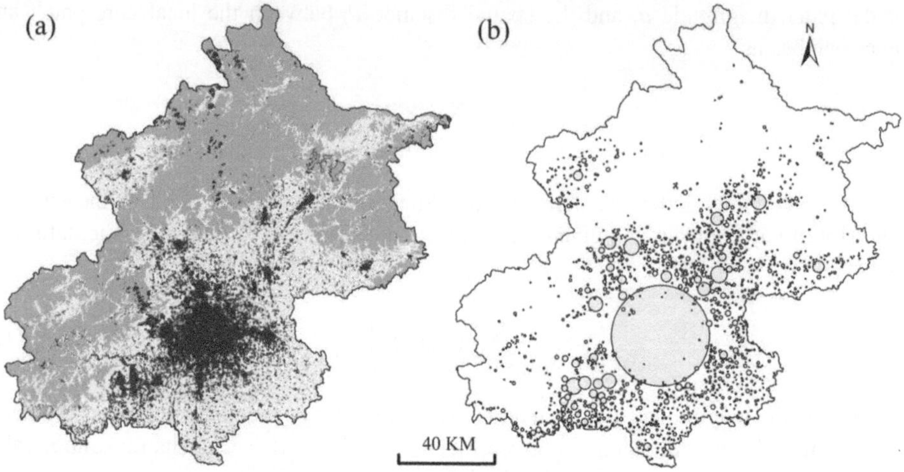

Figure 3. (a) Urban build-up patches (dark areas) in Beijing derived from satellite images in 2004 and (b) corresponding equivalent circles transformed from patch area.

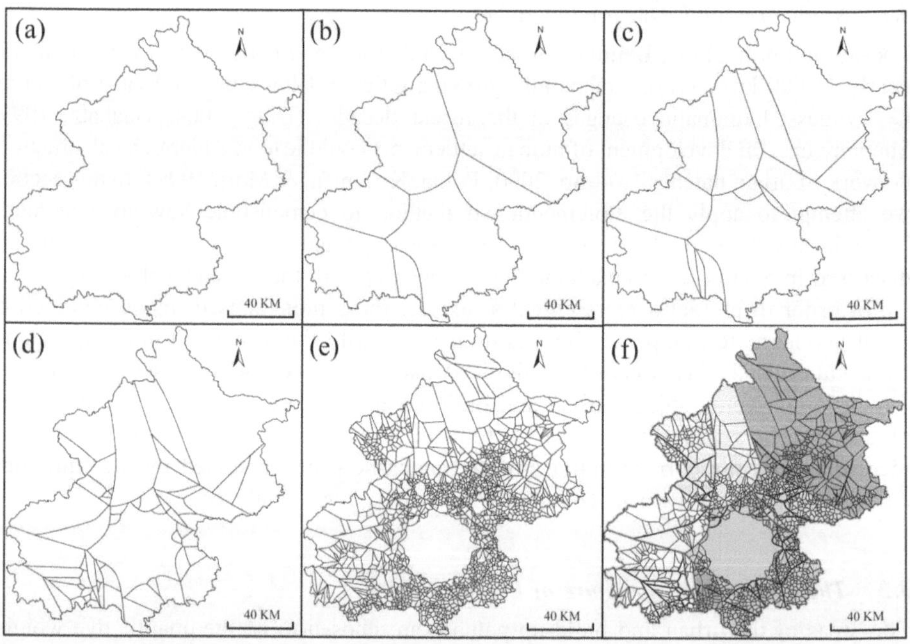

Figure 4. Spatially hierarchical structures. (a) The first-level structure, (b) the second-level structure composed of seven sub-regions, (c) the third-level structure composed of 14 sub-regions, (d) the fourth-level structure composed of 60 sub-regions, (e) the fifth-level structure composed of 1668 sub-regions, and (f) an integration of five-level hierarchical structure.

patches with a total area of 170 km^2. The number of patches are 7, 46, and 1497 for the remaining sub-groups, respectively. Since only one patch object is contained in the first sub-group, the result of the first-order spatial partition is just the entire study area (Figure 4a). The second-order partition based on the patches contained in the first and

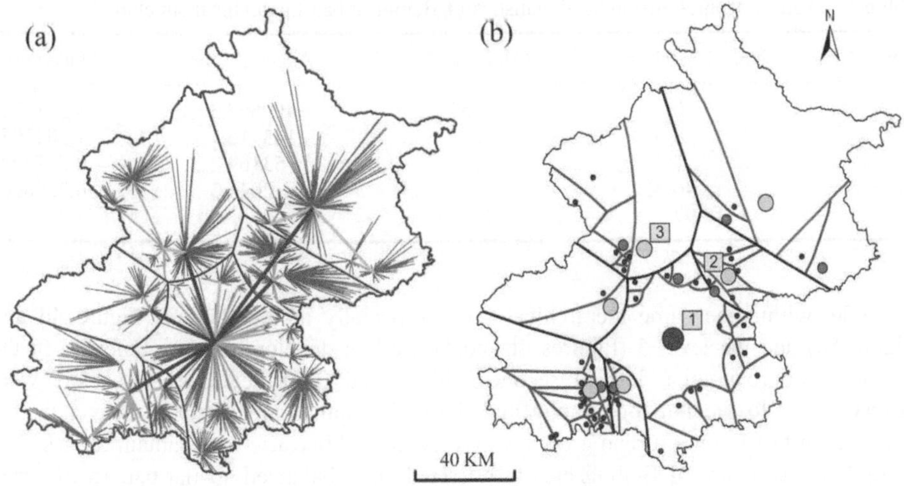

Figure 5. (a) Hierarchical linked network of patches (a thicker line indicates a higher hierarchical level) and (b) four-level hierarchical patches clustered by the natural breaks classification and the corresponding nested sub-regions partitioned by AWV diagram.

second sub-groups produces seven connected Voronoi cells with the consideration of the patch weights (here are the radii of patches) (Figure 4b). The patches contained in the third sub-group are allocated to the relevant Voronoi cells according to the spatial relationships between the patch centriod and the Voronoi cell. In each sub-region, we repeat partitioning all contained patches by using the AWV diagram. A five-level hierarchical structure is constructed by running the spatial partition and allocation operations five times as shown in Figure 4. The hierarchical link network is also established on the basis of the hierarchical regions and grouped patches. The first depth link is defined as the connections between the primary patch (i.e., the patch in the first sub-group) and the patches contained in the second sub-group, which are treated as the second-level local core patches for the second-level sub-regions formed by Voronoi cells. Generally, the kth level local patches are connected to the $(k + 1)$th patches by the kth depth links. The hierarchical link network for urban and build-up patches is shown in Figure 5a. Two patches connected by a link are considered as the neighbor patches for each other.

3.3. *Statistical features*

The values of three hierarchical statistics are summarized in Table 1 In general, Beijing urban hierarchical structure is compact in the upper levels, but very loose near the bottom levels. For instance, the top level had a single patch. The patch number of the second sub-group (6) was approximately similar to that of the third sub-group (7). However, the number of patches in the fourth sub-group noticeably increased (Figure 4d), while the number of patches in the fifth sub-group dramatically swelled (Table 1, Figure 4e, Figure 5b). As demonstrated in Figure 1, the distribution of urban patch numbers over a theoretically hierarchical structure should form a symmetric tree. However, Beijing urban hierarchy exhibited an asymmetric structure. In fact, the spatial distribution of patches reflected the influences of the environment and the history of urban development. For

Table 1. Values of three hierarchical statistics of Beijing urban hierarchical structure.

Level	No. of patches	Connectivity	Accessibility	Uncertainty
1	0	1557	409667.3	3.561972
2	6	509	143916.7	1.502505
3	7	167	53169.2	0.527299
4	46	54	2046.3	0.228666
5	1,497	N/A	N/A	N/A

example, within the same hierarchical level, especially at the level 4 (Figure 4d and Figure 5b) and the level 5 (Figures 4f and 5b), urban development was uneven. In the southern suburban area, lots of second-, third-, and fourth-level urban centers were observed due to the flat topography and the high population density. However, in the northern part of Beijing, urban growth was very limited because of mountain ranges. The hierarchical structure of Beijing clearly reflected an imbalanced spatial pattern of urban development and build-up areas in Beijing.

Moreover, the connectivity and the accessibility of the urban hierarchical tree of Beijing were decreasing from the top to the bottom, which indicated that the interactions between urban development centers were decreasing along with the sizes. On the other hand, the diversity of the urban hierarchical structure was measured by the index of uncertainty. In general, the uncertainty decreased from the top hierarchy toward the bottom hierarchy. In other words, the diversity of patch compositions was diminishing from the top to the bottom. Furthermore, the uncertainty index was also meaningful at a given hierarchical level. For instance, the high uncertainty (e.g., 4.3 for Patch 3 at the third level on Figure 5b) indicated that the patches located in that sub-region and connected to the same local core patch displayed distinct spatial patterns. In other words, there existed obvious spatial disparity of urban growth in that sub-region. For instance, several growth centers were clustered near the west vicinity of Patch 3 (Figure 5b).

4. Conclusions

The hierarchically networked structure of patchy landscape is a representation of self-organization of complex systems involving complicated landscape dynamics in the context of patch evolution, interactions, and environmental influences. Conceptual models derived from the percolation theory show that hierarchical patterns of landscape can be reproduced by random algorithms at multiple-scale tiling regions (Lavorel et al., 1993). However, it remains a challenge to generate a spatially hierarchically networked configuration from increasingly available censored and observed landscape data. It is less practical to use random maps to understand landscape pattern development, spatial heterogeneity, and underlying nonlinear interactions of patches from the urban modeling and planning perspectives.

Recent studies in networked city systems and landscape ecology have inspired this research. The goal of our research is to develop an approach combining both statistical and geometric processes for delineating a hierarchically networked structure from unordered observed data derived from remote sensing. Our method employs natural breaks classification to cluster the initial patch sets into a series of different groups according to a specified geometric or functional attribute. The grouped and ordered patch sets are then spatially portioned by applying the AWV diagram on ArcGIS platform. The hierarchical levels of different patch groups are considered and retained. The patches at a hierarchical level are spatially allocated to the Voronoi cells at the same hierarchical level according to the spatial relationship between the patch centroids and the Voronoi cells. The hierarchical structure is constructed by iterating spatial partition and allocation operations at each hierarchical level. The hierarchical linked network is formed on the basis of the nested hierarchical sub-regions and by linking the patch sets contained in the nested sub-regions following the hierarchical order from top to bottom.

Differing from the theoretical configuration, a realistic hierarchical structure for a given urban and build-up area is often unsymmetrical as a result of spatial heterogeneity. Three statistical indices are developed to provide quantitative descriptions of the hierarchical structure. The structural bias can be evaluated by the index of connectivity. The index of accessibility is used to estimate the impact of neighbor patches to the local core patch. The structural variation is assessed by the index of uncertainty.

The motivation of this research is to develop a systematic method to measure spatial patterns of urban growth graphically and quantitatively across various regions so that a comparative analysis could be conducted to examine dynamic trajectories and spatial characteristics of urban growth at different stages and sizes. Currently, the proposed hierarchically networked structure of urban landscape was tested in Beijing City of China. This case study focused on constructing a hierarchical urban network out of the urban land patches on the basis of the classified land uses from a Landsat image. The case study illustrated that the hierarchical graphic urban framework obtained by the proposed method closely resembled the urban landscape structure of Beijing City and the quantitative indices were effective in describing the spatial pattern of Beijing City.

It should be pointed out that the delineation method for constructing a hierarchical networked structure can be applied in ecological research, urban modeling, and urban ecosystem assessment. The method can be used to examine temporal changes and spatial patterns of other types of land uses or land covers. For example, this method could be potentially used to identify emergent dynamics of desertification of agricultural lands in West China. Partitioning sand dunes into hierarchically ordered groups with consideration of their sizes and constructing links between them based on their spatial relationships would create a hierarchically networked sand dunes graphic structure. Through comparison of the sand dunes networked structures at different times, we could examine the intrusion or recession dynamics of sand dunes. By comparing the sand dunes networked structures in different places, we could detect regional disparity and spatial pattern of sand dunes evolution. A second use of the hierarchical delineation method is to facilitate the assessment of interactions between urban patches and other types of land patches, which could significantly improve prediction accuracy of urban models. For instance, Chen and his colleagues (2014) developed a patch-based cellular automata (CA) model to overcome the lack of interactions between urban growth and environments confronted in the conventional cell-based CA simulation strategy. The spatial pattern similarity test confirmed that the patch-based CA simulation was over 18% higher than that of the cell-based CA. It is logical to suggest that the simulation outcome could be further improved if the interconnections between urban patches and other types of land uses could be modeled at varied hierarchical levels (i.e., at different spatial scales). Another potential application of the hierarchical delineation method is to quantitatively assess ecological functions of urban built infrastructures because urban patches are treated as part of a broader ecological system by the hierarchical delineation approach. For example, Yan and his collaborators (2013) used an agent-based model to simulate the physical and socioeconomic drivers of urban expansion and the impacts on cropland, forest, water area, and grassland under different climate change scenarios. In short, the patch-based ecological approach inherited in the hierarchical structure delineation method facilitates assessments of interconnectedness and interaction between different types of land uses and land covers, which is increasingly seen in newly developed analytical methods in urban modeling, environmental science, geographical information science, and landscape ecology.

However, the case study reported in this article is a prototype. There are many opportunities for more testing and further improvement. Although the Beijing case study successfully depicted the hierarchical networked urban landscape and described the statistical characteristics of its spatial structure, two additional experiments would help provide deeper insights. In other words, a cross-regional comparison between Beijing and another major Chinese city, or a longitudinal analysis of the spatial pattern changes of Beijing in different periods, would tell more about the effectiveness of the graphic framework and the sensitiveness of the quantitative indices. These tests are our next research efforts and important milestone for proving the hierarchical urban landscape structure design.

Second, a robust algorithm of constructing a hierarchically networked urban system should consider "flows and interactions" between urban patches. Although an attribute of urban patches could be used as the weight for determining AWV partitions in the current method, model-based or statistical approach should be introduced in order to investigate the complexity embedded in urban networks. Third, the determination of the number of hierarchical levels in the current method was relying on an operator's domain knowledge and prone to human errors. A rigorous method should be developed on the basis of newly developed data mining techniques, such as similarity measurements (Lhermitte, Verbesselt, Verstraeten, & Coppin, 2011). Fourth, in this study, we used equivalent circles to replace the actual land patches in the partitioning process for reducing the computational complexity. We should conduct further experiments to construct the hierarchically networked structure directly from the original land patches in order to obtain more precise hierarchical information for patchy landscape.

Finally, we want to attribute this research to Professor James Wheeler. The design of the hierarchical networked structure reflects several important contributions of Professor Wheeler to urban-geographic research. Economic activity and social interaction were important flows and interactions observed in urban space (Wheeler, 1973), which contributes to recent acceptance of the concept of cities as networked systems. This notion has inspired our interest in this research. In fact, the hierarchical urban structure delineation method developed in this article is a computational implementation of Wheeler's linked metropolitan structure (Wheeler, 1968). Furthermore, Professor Wheeler was an early adopter of GIS and quantitative analysis in urban geography studies (Wheeler, 1997). The hierarchical networked structure delineation method is an application of GIS for conducting quantitative analysis of urban structures. With this tool, we can more effectively study urban hierarchies and urban growth.

Disclosure statement

No potential conflict of interest was reported by the authors.

References

Amson, John C. (1972). Equilibrium models of cities: 1. An axiomatic theory. *Environment and Planning*, *4*(4), 429–444.

Anselin, Luc, Syabri, Ibnu, & Kho, Youngihn (2006). GeoDa: An introduction to spatial data analysis. *Geographical Analysis*, *38*(1), 5–22.

Antrop, Marc (2000). Background concepts for integrated landscape analysis. *Agriculture, Ecosystems & Environment*, *77*(1–2), 17–28.

Bagan, Hasi, & Yamagata, Yoshiki (2012). Landsat analysis of urban growth: How Tokyo became the world's largest megacity during the last 40 years. *Remote Sensing of Environment*, *127*, 210–222.

Batty, Michael (2013). *The new science of cities*. Cambridge, MA: The MIT Press.

Batty, Michael, & Xie, Yichun (1994). Research article. Modelling inside GIS: Part 1. Model structures, exploratory spatial data analysis and aggregation. *International Journal of Geographical Information Systems*, *8*(3), 291–307.

Batty, Michael, & Xie, Yichun (1996). Preliminary evidence for a theory of the fractal city. *Environment and Planning A*, *28*(10), 1745–1762.

Bogaert, Jan, Farina, Angelo, & Ceulemans, Reinhart (2005). Entropy increase of fragmented habitats: A sign of human impact? *Ecological Indicators*, *5*(3), 207–212.

Chen, Yimin, Li, Xia, Liu, Xiaoping, & Ai, Bin (2014). Modeling urban land-use dynamics in a fast developing city using the modified logistic cellular automaton with a patch-based simulation strategy. *International Journal of Geographical Information Science*, *28*(2), 234–255.

Conzen, Michael P. (1978). Analytical approaches to the urban landscape. In K. W. Butzer (Ed.), *Dimensions in human geography: Essay in some familiar and neglected themes*, Research Paper 186 (pp. 128–165). Chicago, IL: Department of Geography, University of Chicago.

Conzen, Michael R. G. (1960). Alnwick, Northumberland: A study in town-plan analysis. *Transaction and Papers (Institute of British Geographers Publication)*, *27*, iii+ix-xi+1+3–122.

Conzen, Michael R. G., & Whitehand, Jeremy W. R. (1981). *The urban landscape: Historical development and management*. London: Academic Press.

Diestel, Reinhard (2005). *Graph theory*. New York, NY: Springer-Verlag Heidelberg.

Ding, Chengri (2004). Urban spatial development in the land policy reform era: Evidence from Beijing. *Urban Studies*, *41*(10), 1889–1907.

Dong, Pinliang (2008). Generating and updating multiplicatively weighted Voronoi diagrams for point, line and polygon features in GIS. *Computers & Geosciences*, *34*(4), 411–421.

Edelsbrunner, Herbert, & Seidel, Raimund (1986). Voronoi diagrams and arrangements. *Discrete & Computational Geometry*, *1*(1), 25–44.

Eppstein, David (2005). *Geometry in action – A collection of applications of computational geometry*. Retrieved from http://www.ics.uci.edu/~eppstein/geom.html

Forman, Richard T. T., & Godron, Michel (1981). Patches and structural components for a landscape ecology. *Bioscience*, *31*(10), 733–740.

Forman, Richard T. T., & Godron, Michel (1986). *Landscape ecology*. New York, NY: Wiley.

Frank, Karen A., & Schneekloth, Lynda (Eds.). (1994). *Ordering space: Types in architecture and design*. New York, NY: Van Nostrand Reinhold.

Fujita, Masahisa, Krugman, Paul, & Mori, Tomoya (1999). On the evolution of hierarchical urban systems. *European Economic Review*, *43*(2), 209–251.

Gahegan, Mark, & Lee, Ickjai (2000). Data structures and algorithms to support interactive spatial analysis using dynamic Voronoi diagrams. *Computers, Environment and Urban Systems*, *24*(6), 509–537.

Gaubatz, Piper (1999). China's urban transformation: Patterns and processes of morphological change in Beijing, Shanghai and Guangzhou. *Urban Studies*, *36*(9), 1495–1521.

Gaucherel, Cédric, Fleury, David, Auclair, Daniel, & Dreyfus, Philippe (2006). Neutral models for patchy landscapes. *Ecological Modelling*, *197*(1–2), 159–170.

Gómez, Cristina, White, Joanne C., & Wulder, Michael A. (2011). Characterizing the state and processes of change in a dynamic forest environment using hierarchical spatio-temporal segmentation. *Remote Sensing of Environment*, *115*(7), 1665–1679.

Gong, Hong M., & Wheeler, James O. (2002). The location and suburbanization of business and professional services in the Atlanta area. *Growth and Change*, *33*(3), 341–369.

Goovaerts, Pierre (1997). *Geostatistics for natural resources evaluation*. New York, NY: Oxford University Press.

Halls, Peter J., Bulling, Mark, White, Piran C. L., Garland, L., & Harris, Stephen (2001). Dirichlet neighbours: Revisiting Dirichlet tessellation for neighbourhood analysis. *Computers, Environment and Urban Systems*, *25*(1), 105–117.

Handy, Susan L., & Niemeier, Debbie A. (1997). Measuring accessibility: An exploration of issues and alternatives. *Environment and Planning A*, *29*(7), 1175–1194.

Jenks, George F., & Coulson, Michael R. C. (1963). Class intervals for statistical maps. *International Yearbook of Cartography*, *3*, 119–134.

Jorgenson, M. Torre (2000). Hierarchical organization of ecosystems at multiple spatial scales on the Yukon-Kuskokwin Delta, Alaska, USA. *Artic, Antarctic, and Alpine Research*, *32*(3), 221–239.

Karavelas, Menelaos I., & Yvinec, Mariette (2002). Dynamic additively weighted Voronoi diagram in 2D. In Rolf H. Möhring and Rajeev Raman (Eds.), *Proceedings of the 10th annual European Symposium on Algorithms*, Lecture Notes in Computer Science #2461 (pp. 586–598). Berlin: Springer-Verlag.

Kotliar, Natasha B., & Wiens, John A. (1990). Multiple scales of patchiness and patch structure: A hierarchical framework for the study of heterogeneity. *Oikos*, *59*(2), 253–260.

Lavorel, Sandra L., Gardner, Robert H., & O'Neill, Robert V. (1993). Analysis of patterns in hierarchically structured landscapes. *Oikos*, *67*(3), 521–528.

Lhermitte, Stef, Verbesselt, Jan, Verstraeten, Willem W., & Coppin, Pol (2011). A comparison of time series similarity measures for classification and change detection of ecosystem dynamics. *Remote Sensing of Environment*, *115*(12), 3129–3152.

McGarigal, Kevin, & Marks, Barbara J. (1995). *FRAGSTATS: Spatial pattern analysis program for quantifying landscape structure* (USDA Forest Service General Technical Report PNW-351). Portland, OR: USDA Forest Service, Pacific Northwest Research Station.

McMaster, Robert, & McMaster, Susanna (2002). A history of twentieth-century American academic cartography. *Cartography and Geographic Information Science*, *29*(3), 305–321.

Mennis, Jeremy, & Liu, Jun W. (2005). Mining association rules in spatio-temporal data: An analysis of urban socioeconomic and land cover change. *Transactions in GIS*, *9*(1), 5–17.

Moudon, Anne V. (1994). Getting to know the built landscape: Typomorphology. In K. A. Franck & L. H. Schneekloth (Eds.), *Ordering space: Types in architecture and design* (pp. 289–311). New York, NY: Van Nostrand Reinhold.

Moudon, Anne V. (1997). Urban morphology as an emerging interdisciplinary field. *Urban Morphology*, *1*, 3–10.

Muratori, Saverio (1959). *Studi per una operante storia urbana di Venezia*. Rome: Istituto Poligrafico dello Stato.

Neal, Zachary P. (2013). *The connected city – How networks are shaping the modern metropolis and modern life*. New York, NY: Routledge.

Newman, Mark, Barabási, Albert L., & Watts, Duncan J. (2006). *The structure and dynamics of networks*. Princeton, NJ: Princeton University Press.

Nielsen, Donald R., & Wendroth, Ole (2003). *Spatial and temporal statistics: Sampling field soils and their vegetation*. Reiskirchen: Catena Verlag GMBH.

O'Rourke, Joseph (1998). *Computational geometry in C*. New York, NY: Cambridge University Press.

O'Sullivan, David (2000). Using desktop GIS for the investigation of accessibility by public transport: An isochrone approach. *International Journal of Information Science*, *14*(1), 85–104.

Openshaw, Stan, & Veneris, Yannis (2003). Numerical experiments with central place theory and spatial interaction modelling. *Environment and Planning A*, *35*(8), 1389–1403.

Pataki, Diane E., Carreiro, Margaret M., Cherrier, Jennifer, Grulk, Nancy E., Jennings, Viniece, Pincetl, Stephanie, … Zippere, Wayne C. (2011). Coupling biogeochemical cycles in urban environments: Ecosystem services, green solutions, and misconceptions. *Frontiers in Ecology and the Environment*, *9*(1), 27–36.

Perino, Grischa, Andrews, Barnaby, Kontoleon, Andreas, & Bateman, Ian (2014). The value of urban green space in Britain: A methodological framework for spatially referenced benefit transfer. *Environmental and Resource Economics*, *57*(2), 251–272.

Pickett, Steward T. A., & White, Peter S. (1985). *The ecology of natural disturbance and patch dynamics*. New York, NY: Academic Press.

Pijanowski, Bryan C., Tayyebi, Amin, Doucette, Jarrod, Pekin, Burak K., Braun, David, & Plourde, James (2014). A big data urban growth simulation at a national scale: Configuring the GIS and

neural network based land transformation model to run in a high performance computing (HPC) environment. *Environmental Modelling & Software*, *51*(1), 250–268.
Portugali, Juval (2010). *Complexity, cognition and the city*. Heidelberg: Springer.
Rey, Sergio J., & Janikas, Mark V. (2006). STARS: Space-time analysis of regional systems. *Geographical Analysis*, *38*, 67–86.
Sipper, Moshe (1997). *Evolution of parallel cellular machines: The cellular programming approach*. Berlin: Springer.
Sozio, Giulia, Mortelliti, Alessio, Boccacci, Francesca, Ranchelli, Elisa, Battisti, Corrado, & Boitani, Luigi (2013). Conservation of species occupying ephemeral and patchy habitats in agricultural landscapes: The case of the Eurasian reed warbler. *Landscape and Urban Planning*, *119*, 9–19.
Srinivasan, Vijay, & Nackman, Lee R. (1987). Voronoi diagram for multiply-connected polygonal domains I: Algorithm. *IBM Journal of Research and Development*, *31*(3), 361–372.
Szell, Michael (2014). Connecting paradigms. *Science*, *343*, 970–971.
Tao, Sui, Corcoran, Jonathan, Mateo-Babiano, Iderlina, & Rohde, David (2014). Exploring bus rapid transit passenger travel behaviour using big data. *Applied Geography*, *53*, 90–104.
Tian, Guangjin, Yang, Zhifeng, & Xie, Yichun (2007). Detecting spatiotemporal dynamic landscape patterns using remote sensing and the lacunarity index: A case study of Haikou City, China. *Environment and Planning B: Planning and Design*, *34*(3), 556–569.
Wheeler, James O. (1968). *Metropolitan linkage-structure: A factor analysis approach*. Akron, OH: East Lakes Division, Association of American Geographers.
Wheeler, James O. (1973). Social interaction and urban space. In Melvin Albaum (Ed.), *Geography and contemporary issues* (pp. 303–308). New York, NY: John Wiley & Sons. (reprinted from *Journal of Geography*).
Wheeler, James O. (1997). Quantitative analysis and GIS in urban geography: Then and now. *Urban Geography*, *18*(2), 95–99.
Whitehand, Jeremy W. R. (1992). Recent advances in urban morphology. *Urban Studies*, *29*(3–4), 619–636.
Whitehand, Jeremy W. R., & Larkham, Peter J. (Eds.). (1992). *Urban landscapes: International perspectives*. London: Routledge.
Wu, Jianguo (2007). *Landscape ecology: Pattern, process, scale and hierarchy* (2nd ed.). Beijing: Higher Education Press. [in Chinese].
Wu, Jianguo, & David, John L. (2002). A spatially explicit hierarchical approach to modeling complex ecological systems: Theory and applications. *Ecological Modelling*, *153*(1–2), 7–26.
Wu, Jianguo, & Loucks, Orie L. (1995). From balance of nature to hierarchical patch dynamics: A paradigm shift in ecology. *The Quarterly Review of Biology*, *70*(4), 439–466.
Xie, Yichun, Batty, Michael, & Zhao, Kang (2007). Simulating emergent urban form using agent-based modeling: Desakota in the Suzhou-Wuxian region in China. *Annals of the Association of American Geographers*, *97*(3), 477–495.
Xie, Yichun, Yu, Mei, Tian, Guangjin, & Xing, Xuerong (2005). Socio-economic driving forces of arable land conversion: A case study of Wuxian City, China. *Global Environmental Change*, *15*(3), 238–252.
Xie, Yichun, Sha, Zongyao, & Bai, Yongfei (2010). Classifying historical remotely sensed imagery using a tempo-spatial feature evolution (T-SFE) model. *ISPRS Journal of Photogrammetry and Remote Sensing*, *65*(2), 182–190.
Xie, Yichun, & Ye, Xinyue (2007). Comparative tempo-spatial pattern analysis: CTSPA. *International Journal of Geographical Information Science*, *21*(1), 49–69.
Xie, Yichun, Yu, Mei, Bai, Yongfei, & Xing, Xuerong (2006). Ecological analysis of an emerging urban landscape pattern – Desakota: A case study in Suzhou, China. *Landscape Ecology*, *21*(8), 1297–1309.
Yan, Dan, Schneider, Uwe A., Schmid, Erwin, Huang, He Q., Pan, Lihu, & Dilly, Oliver (2013). Interactions between land use change, regional development, and climate change in the Poyang Lake district from 1985 to 2035. *Agricultural Systems*, *119*(1), 10–21.
Zhang, Chi, Wu, Jianguo, Grimm, Nancy B., McHale, Melissa, & Buyantuyev, Alexander (2013). A hierarchical patch mosaic ecosystem model for urban landscapes: Model development and evaluation. *Ecological Modelling*, *250*, 81–100.

Index

Note: Page numbers in **bold** type refer to figures
Page numbers in *italic* type refer to tables
Page numbers followed by 'n' refer to notes

For Product Safety Concerns and Information please contact our
EU representative GPSR@taylorandfrancis.com Taylor & Francis
Verlag GmbH, Kaufingerstraße 24, 80331 München, Germany